Farm Animals in Biomedical Research

Edited by

Vladimir Pliška and Gerald Stranzinger

Department of Animal Science
Swiss Federal Institute of Technology (ETH)
Zürich

Contributors:

Anton C. Beynen, Kurt Bürki, Rolf Claus, Yves Claustre, Magali D'Angio, Jaroslav Dedek, Martin Dehnhard, Peter Driscoll, Nathan S. Fechheimer, Jürg J. Hari, Johein Harmeyer, Judith Hasler-Rapacz, Robert Hediger, Reinhard Kaune, Elisabeth Lang, Sandor Lazary, Gerard P. McGregor, Stefan Neuenschwander, Ernst Peterhans, Vladimir Pliška, Jan Rapacz, Gaby Ruff, Bernard Scatton, Wolf Stegmaier, Gerald Stranzinger, Ursula Süss, Karlheinz Voigt, Reto Zanoni

With 56 figures and 18 tables

1990

Verlag Paul Parey · Hamburg und Berlin

Advances in Animal Breeding and Genetics
Supplements to Journal of Animal Breeding and Genetics

Fortschritte der Tierzüchtung und Züchtungsbiologie
Beihefte zur Zeitschrift für Tierzüchtung und Züchtungsbiologie

Edited by / Herausgegeben von F. PIRCHNER, Weihenstephan - R.D. POLITIEK, Wageningen - L.R. SCHAEFFER, Guelph - R. WASSMUTH, Gießen - J.H. WENIGER, Berlin
Executive Editor / Schriftleitung: F. PIRCHNER, Weihenstephan

Issue/Heft 5

CIP-Titelaufnahme der Deutschen Bibliothek

Farm Animals in Biomedical Research / ed. by Vladimir Pliška and Gerald Stranzinger. Contributors: Anton C. Beynen ... - Hamburg; Berlin : Parey, 1990
  (Fortschritte der Tierzüchtung und Züchtungsbiologie ; H. 5)
  ISBN 3-490-30915-4
NE: Pliška, Vladimir [Hrsg.]; Beynen, Anton C. [Mitverf.]; GT

This work is subject to copyright. All rights are reserved, whether the whole or part of the material is concerned, specially those right of translation, reprinting, re-use of illustrations, recitation, broadcasting, reproduction on microfilms or in other ways, and storage in data banks. Duplication of this publication or parts of thereof is only permitted under provisions of the German Copyright Law of September 9, 1965, in its version of June 24, 1985, and a copyright fee must always be paid. Violations fall under the prosecution act of the German Copyright Law. © 1990 Verlag Paul Parey, Hamburg and Berlin. Addresses: Spitalerstrasse 12, D-2000 Hamburg 1; Lindenstrasse 44-47, D-1000 Berlin 61. Printed in Germany by Rasch Druckerei und Verlag GmbH & Co. KG, Bramsche/Osnabrück. Camera-ready manuscript prepared by the Department of Animal Science, Swiss Federal Institute of Technology, CH-8092 Zürich; layout designed by J. Heiniger. Cover design: Jan Buchholz and Reni Hinsch, Hamburg.

ISSN: 0936-8353 · ISBN 3-490-30915-4

Preface

Animal experimentation has always been a subject of public and scientific controversy. Questions concerning its humanistic side on one hand and its methodological problems (e.g., adequacy, suitability of extrapolations to differing conditions or species) on the other hand, were significant in the past and are still before us.

Whereas the first of the two questions undoubtedly affects our whole society, the methodology in animal experiments remains an intrinsic problem of biological and perhaps also other scientific disciplines. As for biomedical research, the main task of animal experiments rests principally on providing data applicable to human biology and medicine. Leaving aside the problem of a data transfer from animals to humans (we touch on this subject in the introductory chapter), a question can certainly arise as to whether it is reasonable to employ farm animals in addition - or as an alternative - to laboratory animals. We are convinced that the answer is yes, and this publication presents a selection of cases which support this opinion. Many circumstances speak in favor of this idea, particularly when a disease model is needed. It is not usual, and even less desirable, to undertake any surgical or pharmacological treatment in order to produce an "artificial" disease model with a farm animal - something which unfortunately is still necessary and common in laboratory species. Such procedures are usually connected with at least some suffering for the animal. However, genetic aberrations in an animal frequently create a condition similar to that known in human pathology. In such a case the animal model can be used to investigate hereditary patterns, pathophysiology, etiology *etc.* of that particular disease in more detail than is possible in the human, or to develop suitable therapeutic procedures. Genetically deviating animals usually suffer much less than those artificially treated since they can spontaneously adapt to their condition. To trace them in a population frequently appears easier for farm animals (flocks are closely watched by persons taking care of them) and moreover, their treatment offers as a rule less problems due to their size. The authors and editors of this book do not naturally enforce the notion of farm animals as being in all instances preferable to laboratory animals as biomedical models. However, since models of many diseases that one would urgently need are still missing, this extension may help.

The main body of contributions published in this volume was presented at the symposium of the Swiss Society of Genetics on *Genetic variants of farm animals as biomedical models* held in Lausanne, Switzerland, on October 7, 1988. The intention of the organizers was not to treat the problem in an exhaustive manner, but rather to point out the value of farm animals as models. Features were demonstrated in several selected examples including metabolism, stress, behavior, and various disease conditions. Maintenance and reproduction of anomalous animals and problems of using animals as models formed a general frame for these examples. The same layout was used for this publication to which a number of others who did not participate in the symposium agreed to contribute. Since these models are still not broadly employed, the comparison between farm animals, laboratory animals, and in many cases humans as well, was emphasized intentionally. In some instances experience with laboratory animals can be employed to establish a new, not yet existent farm animal model; some of the contributions focus attention on such a possibility. This is the reason why a review on transgenic laboratory animals has been included in this book. The possibility that similar techniques will be applied in the future to domestic animals, provided that the very serious ethical and humanistic concerns are addressed first, should not be denied *a priori*.

It should also be mentioned that a comprehensive handbook on this area, comprising short but rather complete descriptions of many existing animal disease models, has been edited by E.J. ANDREWS, B.C. WARD and N.H. ALTMAN (*"Spontaneous Animal Models of Human Disease"*, Vol. I & II, Academic Press, New York, 1979).

The editors are indebted to the many persons who participated in the preparation of this monograph, and foremost to the authors of the contributions. The symposium and the publication were supported by the following organizations and companies:

- Ciba-Geigy AG, Basel, Switzerland
- ETH Zürich (Swiss Federal Institute of Technology), Switzerland
- F. Hoffmann-La Roche & Co. AG, Basel, Switzerland
- H. Wilhelm Schaumann Stiftung zur Förderung der Agrarwissenschaften, Hamburg, F.R.G.
- Laur-Fonds, ETH Zürich, Switzerland
- Lonza AG, Basel, Switzerland
- Opopharma AG, Zürich, Switzerland
- RCC Research and Consulting Company AG, Itingen, Switzerland
- Sandoz AG, Basel, Switzerland.

Our particular thanks are due to all responsible persons within these institutions who arranged for the grants.

And last but not least, the coworkers of the two editors at the ETH in Zürich, especially Mr. JÖRG HEINIGER, Ms. ELISABETH LANG, Dr. VALERIE MADISON and Mr. STEFAN NEUENSCHWANDER undertook with a great deal of enthusiasm the preparation of the manuscript. Dr. PETER DRISCOLL edited the style and language of many of the articles. All this help is gratefully acknowledged.

Zürich, January 1990

V. Pliška
G. Stranzinger

Table of Contents

Preface 3

Animal models

Animal models as tools for biological research
V. Pliška 9

Gene and chromosome homologies in man and other mammals
G. F. Stranzinger and R. Hediger 17

Physiology and genetics

Farm animals as models of cholesterol metabolism
A. C. Beynen 33

The domestic chicken (Gallus domesticus) as an organism for the study of chromosomal aberrations
N. S. Fechheimer 43

Stress and behavior

Stress: biological and biomedical considerations
V. Pliška 57

Stress in domestic pigs
J. J. Hari and E. Lang 69

Novel variants of the stress hormone ACTH in pigs: genetic and pathophysiological considerations.
K. Voigt, W. Stegmaier and G. McGregor 81

Catecholamine determination in pig peripheral plasma as an alternative parameter to characterize stress situations
M. Dehnhard and R. Claus 89

A genetically-based model for divergent stress responses: behavioral, neurochemical and hormonal aspects
P. Driscoll, J. Dedek, M. D'Angio, Y. Claustre and B. Scatton 97

Disease models

Two unique animal models for the study of human metabolic bone diseases
J. Harmeyer and R. Kaune 111

The pig and its plasma lipoprotein polymorphism in studies of atherosclerosis
J. Rapacz and J. Hasler-Rapacz 131

Caprine arthritis encephalitis
E. Peterhans, R. Zanoni, G. Ruff and S. Lazary 147

Hemophilia in sheep and the use of sheep in blood coagulation research
S. NEUENSCHWANDER and V. PLIŠKA 155

Maintainance of anomalous animal lines

Reproductive biotechnology for preserving genetic anomalies in farm animals
U. SÜSS 167

Transgenic animals as models in biomedical research
K. BÜRKI 177

Index 183

Animal models

Animal models as tools for biological research

V. PLIŠKA

This introduction, and most of the chapters of this monograph, deal with animals which carry an inborn deviation from a physiological norm. This status may result in a rather manifold condition of the organism: it may, for instance, lead to clear-cut pathological changes, or modify its environmental responses (stress), change sensitivity to infections, etc. Such animals may no doubt become valuable models of similar conditions that occur in other animal species, including man.

Laboratory *versus* farm animals

Models of this kind have been long known among laboratory animals (MELBY & ALTMAN 1974, ANDREWS et al. 1979, SHIRE 1979). Genetic lines like the nude mouse (PENTELOURIS 1968, RYGAARD & POOLSEN 1976), Brattleboro rat (SOKOL & VALTIN 1982), Zucker hyperlipoproteinemic rat (ZUCKER 1965), spontaneous hypertensive rat (OKAMOTO 1969) and others are good examples. Small laboratory rodents possess many advantages for biomedical investigations: they can be easily bred, controlled, their gestation time and reproductive cycles are short. In spite of this, there may be some good reasons to extend these well established models by including larger domestic animals.

Firstly, some specific types of genetic deviations, important for human medicine and biology, have been recognized only in domestic animals. New, potentially interesting deviations are likely to be discovered since farm animals are closely watched by both breeders and veterinarians. At least in some instances, the anomaly may be preserved for future use. Secondly, the treatment and observation of larger animals are, as a rule, quite easy. Blood sampling, biopsies, electrophysiologic and other measurements are less difficult than in rats or mice. The optimal size of the animal depends naturally upon the purpose of the experiment and the kind of desired treatment. Thirdly, certain legal and ethical aspects are important. In many countries, a veterinarian must be involved in the experiments with domestic animals, even if minor surgery is required, and treatment must follow legally permitted procedures. Also, the "experimental animal" can often stay with the rest of the animal stock and is supposed to follow the daily regime of the farm. Its social isolation from other animals can, in this way, be prevented. And finally, experience has shown that relations between humans and domestic - farm or pet - animals are generally closer than to a laboratory animal. This gives, by itself, the best guarantee that the animal will be treated in the best possible way at any stage of the experiment.

Disease models

Animals with inborn defects frequently mimic a pathological anomaly and are therefore important as disease models. Disease models are commonly employed to develop a new therapeutic procedure or to study, in detail, a disease condition. They are often needed at a certain stage of testing when replacement by a healthy individual is not conceivable: an antiinflammatory drug can scarcely be tested in the absence of inflammation, and - as an extreme example - antibiotics cannot suppress *in vivo* proliferation of bacteria in a noninfected organism. Most of these models, however, are prepared artificially, by a pathogenic treatment of the animal. Common procedures used to these aims are summarized in Table 1. They are, as a rule, highly questionable since they undoubtedly cause considerable pain and suffering to the animal.

An inborn deviation from a biological norm may also act as a pathogenic agent in bringing about various phenotypic changes. In some instances, this condition is identical with a genetic disease known in humans. In some others, it at least mimics symptoms of a human disorder. In most of the cases, however, such an animal fulfills requirements of a disease model and has the great advantage of being better adapted to the pathologic condition than the artificially treated animals. The animal suffers much less and, leaving aside ethical aspects, additional sources of problems in animal experimentation, such as pain and stress, can often be successfully avoided.

Table 1. *Artificial animal disease models (common examples).*

treatment	method
surgical	removal of organs and tissues
	blockade of outlets of glands
	manipulations on blood vessels: blockade
	by-passing
	cross-circulation
	catheterization
	insertion of cannulas
	artificial (surgical) fistulae
physical/mechanical	dynamic stress (Nobel-Collip drum[a])
	irradiation (ionizing, UV, etc.)
chemical	DOCA[b]: induced hypertension in rats
	experimental diabetes mellitus: specific destruction of pancreatic β-cells by alloxan
	models of inflammation: injection of porcelain clay etc.
	cancerogens: tumor induction
	active/passive immunization against endogenous substances
biological	viral, microbial, fungal etc. infections
	gene transfer
	embryo manipulations

[a]*No more in use for ethical and/or legal reasons.*
[b]*Desoxycorticosterone acetate.*

Deficiencies in regulatory systems

Brattleboro and spontaneous hypertensive (SH) rats can be offered as examples of inborn genetic disorders of regulatory pathways. In the first example, the strain displays a spontaneous hypothalamic diabetes insipidus (VALTIN & SCHROEDER 1964, SOKOL & VALTIN 1982) due to an inborn single point mutation of the vasopressin precursor in magnocellular neurosecretory neurons (SCHMALE & RICHTER 1984). As a consequence, the rats are in a permanent state of water diuresis in which the urine osmolarity sinks below 300 mOsm/l and reaches sometimes values as low as 50 mOsm/l. A similar human condition (diabetes insipidus centralis; for review cf. CZERNICHOW & ROBINSON 1985) is known to be inherited in about fifty per cent of manifested cases, the rest being caused by midbrain lesions, traumatic processes in the hypothalamus and hypophysis, or by tumors. The severe hereditary form occurs with a frequency of 1/50,000 to 1/20,000 (homozygotes), but milder forms (heterozygotes) are more frequent. The defect in the Brattleboro rat is inherited in an autosomal-recessive way. Except for a massive polyuria (urine volume increases up to twenty-fold of the norm, and water intake sometimes corresponds to three quarters of the animal's body weight per day) and a slight retardation of growth, the affected homozygous animals show no trace of suffering. They have contributed to two considerable scientific achievements. Firstly, they opened the path to the elucidation of the etiology of this rare but troublesome disease - something which cannot be done by artificial models. Secondly, they played an important role in the development of several new peptide drugs for substitution therapy in this disease, in particular deamino-D-arginine-vasopressin (known as DDAVP). Alternative artificial models for this second purpose, i.e. rats with surgically removed hypophyses (hypophysectomy), rats with electric lesions in the supraoptic region of the brain, or superhydrated and surgically treated rats (catheterization and removing of urinary bladder, cannulation of a jugular vein) with a large, continuously administered intragastric water load, would nowadays be accepted only reluctantly.

The SH-rat, on the other hand, is an animal model which has been developed by a systematic selection of rats in which a sodium-rich diet leads to an increase of arterial blood pressure (OKAMOTO 1969, SEN et al. 1972). Their significance for investigation of the pathogenesis and the therapy of malignant hypertension is undisputed. Artificial models - desoxycorticosterone acetate (DOCA) treatment of rats or surgical denervation of the kidney in dogs - may be suitable for antihypertensive drug design but are of limited use for any other purposes.

These two examples clearly show that a genomic change may in some instances have simple, and in some others very complex, phenotypic consequences. Absence of a hormone is rather frequent. Some other cases of inborn errors in endocrine systems are caused by missing a receptor component, or components of cellular systems transducting hormone signals ("second messenger" mechanisms). Failure of synthesis of a gene product (one or several) is a common denominator of all of them. However, the inverse case, an overproduction, is also conceivable, although still rather unique at present. The so-called NDI-rat (nephrogenic diabetes insipidus rat) may serve as an example of this (FALCONER et al. 1964, NAIK & VALTIN 1969). This strain displays an inborn error of the urine concentrating mechanism, insensitive to vasopressin, akin to the human diabetes insipidus renalis. The inheritance mode in the human is X-chromosomal - recessive, and nonhereditary forms are frequently associated with various tubulonephropathies or disturbances of electrolyte metabolism, for instance in sickle cell anemia. The error may occur in a process subsequent to the initiation of the vasopressin response on receptors in collecting ducts, and may most likely affect the production of the corresponding "second messenger", the cyclic adenosine monophosphate (cAMP), or of some of the protein kinases involved in eliciting the vasopressin effect. Recently, DOUSA and coworkers (1988) identified an excessive activity of one of the intacellular enzymes that control, by inactivation, intracellular levels of cAMP: the cAMP-phosphodiesterase type III (Fig. 1). Thus, the rapid inactivation of the second messenger prevents its intracellular accumulation and leads to the failure of the vasopressin response. This finding will certainly help to elucidate the etiology of the human disease, and hopefully also to facilitate attempts toward its therapeutic control, still very difficult at present.

Enzyme and metabolic defects

A great variety of inborn errors concern genetic errors in enzyme or coenzyme biosynthesis. Similar to other disease conditions, most of them cannot be artificially produced and the only choice is a spontaneous model, if this exists. Again, two examples can be mentioned in this context, both concerning farm animals. The first one is hemophilia (type A or B) which has been diagnosed in several domestic species like the dog, sheep, cat and horse. Missing an enzymatic step in the blood clotting cascade (the "antihemophilic" factor VIII:C or the factor IX) results in serious defects of bleeding control. Employment of the model and problems associated with it are thoroughly discussed in one of the present chapters (NEUENSCHWANDER & PLIŠKA, this monograph).

The other example is the syndrome of malignant hyperthermia, called "halothane reaction", in domestic pigs (EIKELENBOOM & MINKEMA 1974). This condition is especially frequent in the Piétrain breed from Belgium, and practically nonexistent in Duroc pigs. There seems to be a striking similarity between malignant hyperthermia in pigs and in humans. The syndrome counts in human medicine among the fearful post-surgery complications with a high mortality rate (30-40%), occurring in the frequency of about 1:50,000 in surgery patients anesthetized with certain types of general anesthetics, or treated with certain muscle relaxants (estimates taken from SENGUPTA 1980). It is most likely an inborn error of intracellular calcium metabolism with unknown hereditary pattern. In pigs, on the other hand, the genetic background has been well recognized: the condition is caused by a mutation of apparently a single gene localized on chromosome 6 (DAVIES et al. 1988), within a known coupling group (VÖGELI et al. 1984). The inheritance is recessive, with the gene showing a 100% penetrance.

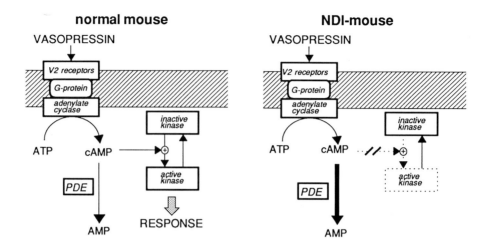

Fig. 1. *Inborn defect in the collecting duct of the kidney of nephrogenic diabetes insipidus (NDI) mouse. Hatched area symbolizes the epithelial cell membrane carrying vasopressin receptors (type V2) which regulate the water flux from the lumen of the duct to the interstitial space. Excessive activity of the cAMP-phosphodiesterase type III (PDE) strongly reduces the intracellular level of cAMP which acts as a second messenger of vasopressin. Insufficient activation of the protein kinase which controls the cell membrane permeability for water causes an inhibition of the water flux. (According to DOUSA et al. 1988).*

A number of chromosomal serum markers, which enable a safe prediction regarding the halothane reaction, are known. Using halothane positive pigs as a model of the human condition (HARRISON 1975), progress in the therapeutic control of manifest malignant hyperthermia has been recently achieved by introducing a muscle relaxant drug with calcium stabilizing effects (Dantrolene sodium). The model may very considerably facilitate the development of these life-saving drugs in the future. At the same time, we are faced here with the problem of preserving a deviating animal line for future research. The "halothane gene" has been systematically eliminated from the stocks of afflicted breeds by suitable breeding strategies. This is indeed understandable from an economic point of view: high mortality and sensitivity to stress of halothane positive animals cause sometimes a considerable loss of animals. At the same time, however, valuable animal material which may be urgently needed for biomedical purposes in the near future gets inevitably lost. A way out of this dilemma may perhaps rest in new reproductive techniques which allow the preservation of embryos or sperm and, therefore, offer possibilities to renew breeding at a later time (cf. SÜSS, this monograph).

Anatomic deviations

Finally, a number of macroanatomic defects which resemble those in humans have been found in domestic animals. Their hereditary pathway is usually unknown but they systematically occur in single animal families (both maternal and paternal lines). Two such deviations in sheep, both described by STRANZINGER et al. (unpublished), can be given as examples. The first one is the inborn inward curling of the eyelid known to human pathology as entropium congenitum. Fig. 2 shows an analogous condition in a 2 month old lamb (White Alpine race of sheep). The other example is the congenital palate deformation (brachygnathia, micrognathia) in sheep of the same race (Fig. 3). Again, there is at least some similarity to the Pierre-Robin syndrome in newborn children (deformation of the palate cleft) - a serious condition which requires an instant surgical treatment. A similar deformation has been observed in the cattle (STRANZINGER et al., unpublished).

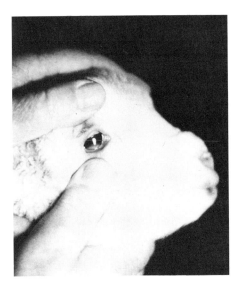 

Fig. 2. Congenital entropium in a 2 month old female lamb (White Alpine race of sheep). Left: manifest disease; right: normal control. (STRANZINGER et al., unpublished).

The problem of species differences and extrapolations

A frequent objection against animal models is their alleged insufficiency to mimic human conditions, owing to species differences. Anatomical, physiological, biochemical, genetical, ethological, ecological and other divergencies are undisputable facts and may prove, upon occasion, prohibitive to the introduction of a particular model. On the other hand, many features of living systems display an amazing similarity among species which are sufficiently close from an evolutionary standpoint. Therefore, the chances for a correct extrapolation of biological findings from one species to another often depend upon, first, the phyletic distance and second, the phenetic relationships (observable phenotypic similarity) of the two species in question.

A global rejection of animal models by many is commonly based on a misunderstanding of the functions of models as methodological supports in testing working hypotheses. Models were in the cradle of modern science; scientists have used them since the era of NEWTON and KEPLER, if not earlier, and their employment has enabled them to understand many complex natural phenomena. Any scientific discipline is unthinkable without them today, and biomedical research is no exception. Animal models are not only semantically similar to other models of natural phenomena; they are based on the same general presumptions and are not even considerably more complex than those of some nonliving physical systems.

Models and similarity

Features of a model

A model and an original are systems of certain properties. A system, in general, is a set composed of two subsets: elements and relationships between them. The nature of the elements and of their relationships may be very different; parts of an engine, cells or subcellular particles of an organism, molecules in a crystal lattice, elements af a mathematical structure (e.g., natural numbers), semantic description of a behavioral pattern etc., can be treated as systems in the same way, using the same axioms. Models can therefore be described in categories of the systems theory and subjected to criteria

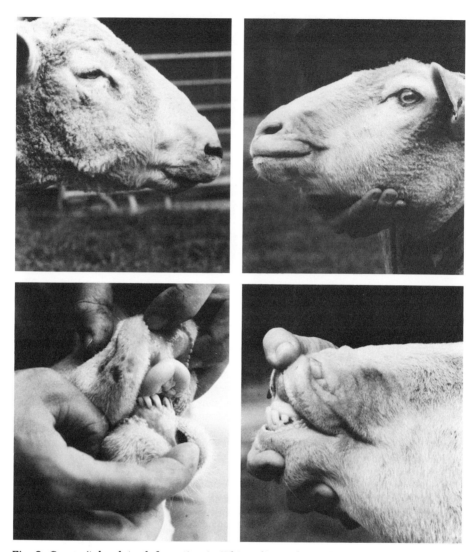

Fig. 3. Congenital palate deformation in White Alpine Sheep (female, age: 2 years). Left: manifest disease; right: normal control. Note the divergent positions of teeth relative to the palate in normal and diseased animals (lower row).

of the systems analysis. For a "less complex" model this is a clear-cut advantage: systems analysis has recently become a well developed discipline, possessing methodological tools that allow largely objective tests of various hypotheses (EYKHOFF 1974). When used in biological and similar systems of high complexity, systems analysis may frequently appear to be somewhat meticulous. Nevertheless, the employment of such a strong set of criteria would certainly be a great advantage, also for this class of models.

Models can be virtually classified as conceptual (phenomenological), mathematical (analytical) or physical (empirical). Their nature may be rather manifold, but their relationships to the originals are stated by three general axioms (LAUE 1971):

Mapping axiom: a model is a system that reflects the features of a natural or artificial original; it mimics elements of the original and/or relationships between them. (An original can itself be a model).

Reduction axiom: a model represents only a defined set of features of the original; the original and its models are therefore similar, but not identical. A model which would mimic *all* features of the original would have to reflect all its elements and all relations between them; such a model would become identical with its original.

Subjectivity axiom: the representative function of a model applies solely to a defined set of subjects, and is restricted to certain operations within a certain domain and a certain time interval.

Biomedical models: their nature

As for biomedical models, similarity criteria are usually related to input-output behavior, i.e., stimulus/response relationships in the model have to be analogous to those of the original, notwithstanding the possible structural differences between the two systems (structure is treated as a "black box"). In other instances, analogy of structural properties may be required (steric models of macromolecules, cell or virus organization, anatomical models, etc.). However, also biomedical models reflect only selected features of the original, thus enabling the investigation of only a certain group of structural or behavioral patterns.

The relevance of a biological model and its performance are assessed very differently by individual scientists. The value of a model may be coarsely overestimated when the knowledge collected is used to reconstruct the original in too much detail. On the other hand, its use may be fully denied by some, who may argue against any possibility of extrapolation, whatsoever. The argument is frequently met in studies of therapeutic procedures; the strategy employed in this area may at best elucidate the problem of modelling in biomedicine. Usually, an animal model can clarify the general mode of action of a drug, indicate after-effects of a physical treatment like irradiation or changes in barometric pressure. If close enough to humans (in the phyletic sense), it may yield even more detailed information, for instance with regard to the dosage of a drug. More certainty concerning the effects on humans can be gained in experiments on humans alone, but even this information is lacking generality: in any particular therapeutic case, the specific patient may react differently from the general scheme, due to very manifold circumstances. Eventually, each patient becomes a model on which the employed therapeutic procedure is tested, with an *a priori* unknown, but usually fairly predictable, outcome. The knowledge obtained with such a "model" is rationally or instinctively used by the physician for treatment of his other patients. An animal - or other biological - model is a link in this deductive process and it would be fully illogical to negate its considerable significance. The same holds true for other aims for which animal models are employed (e.g., etiology, or heredity, of a disease).

Clear specification of the goal which should be reached by employment of the model is always a good starting point for successful deduction, and for a satisfactory solution, of any research problem.

Conclusions

Employment of analogy and notions based on models are firm attributes of contemporary scientific methodology, and only one very simple point should be re-emphasized in conclusion: animal models play an important role in biomedical research, and no forms of reasoning, be they epistemological, methodological, ethical (in the common understanding of ethics) or other, can convincingly argue against their rational employment. We must remain aware, however, that many queries concerning animal experimentation are presently waiting for satisfactory solutions.

Acknowledgements

Critical comments by Professor G. STRANZINGER and Dr. P. DRISCOLL helped to discover many weak points of the author's argumentation used in this article. This collegial help is gratefully acknowledged.

References

ANDREWS, E.J.; WARD, B.C.; ALTMAN, N.H. (Editors) (1979): Spontaneous Animal Models of Human Diseases, Vol. 1, 2. Academic Press, New York, NY.

CZERNICHOW, P.; ROBINSON, A.G. (Editors) (1985): Diabetes insipidus in Man. S. Karger, Basel.

DAVIES, W.; HARBITZ, I.; FRIES, R.; STRANZINGER, G.; HAUGE, J.G. (1988): Porcine malignant hyperthermia carrier detection and chromosomal assignment using a linked probe. Anim. Genet. 19, 203-212.

DOUSA, T.P.; HOMMA, S.; VALTIN, H.; COFFEY, A. (1988): Cellular pathogenesis of nephrogenic diabetes insipidus in mice. In: Recent Progress in Posterior Pituitary Hormones 1988 (Editors: YOSHIDA, S. & SHARE, L.). Excerpta Medica International Congress Series No. 797. Excerpta Medica, Amsterdam. Pp. 321- 326.

EIKELENBOOM, G.; MINKEMA, D. (1974): Prediction of pale, soft, exudative muscle with non-lethal test for the halothane-induced porcine malignant hyperthermia syndrome. Netherland's J. Vet. Sci. 99, 421-426.

EYKHOFF, P. (1974): System Identification. Parameter and State Estimation. J. Wiley & Sons, Inc., London. Pp. 1-9.

FALCONER, D.S.; LATYSZEWSKI, M.; ISAACSON, J.H. (1964): Diabetes insipidus associated with oligosyndactyly in the mouse. Genet. Res. 5, 473-488

HARRISON, G.G. (1975): Control of the malignant hyperpyrexic syndrome in MSH (malignant hyperpyrexic susceptible) swine by dantrolene sodium. Brit. J. Anaesth. 47, 62-65.

LAUE, R. (1971): Elemente der Graphentheorie und ihre Anwendung in den biologischen Wissenschaften. F. Vieweg & Sohn GmbH, Braunschweig. Pp. 16-22.

MELBY, E.C., Jr.; ALTMAN, N.H. (Editors) (1974): Handbook of Laboratory Animal Science, Vol. 2. CRC Press, Inc., Cleveland, OH.

NAIK, D.V.; VALTIN, H. (1969): Hereditary vasopressin-resistant urinary concentrating defect in mice. Amer. J. Physiol. 217, 1183-1190.

OKAMOTO, K. (1969): Spontaneous hypertension in rats. Int. Rev. Exp. Pathol. 7, 227-270.

PENTELOURIS, E.M. (1968): Absence of thymus in a mutant mouse. Nature 217, 370-371.

RYGAARD, J.; POOLSEN, C.O. (1976): The nude mouse: the hypothesis of immunological surveillance. Transplant. Rev. 28, 43-63.

SCHMALE, H.; RICHTER, D. (1984): Single base deletion in the vasopressin gene is the cause of diabetes insipidus in Brattleboro rats. Nature 308, 705-709

SEN, S.; HOFFMAN, G.C.; STOWE, N.T.; SMEBY, R.R.; BUMPUS, F.M. (1972): Spontaneous hypertension and erythrocytosis in rats. In: Spontaneous Hypertension. Its Pathogenesis and Complications (Editor: OKAMOTO, K.). Springer Verlag, Berlin. Pp. 227-230.

SENGUPTA, C. (1980): Malignant Hyperthermia - A Biochemical Study. Thesis No. 6718, ETH Zürich.

SHIRE, J.G.M. (Editor) (1979): Genetic Variation in Hormone Systems, Vol. 1, 2. CRC Press, Inc., Boca Raton, FL.

SOKOL, H.W.; VALTIN, H. (Editors) (1982): The Brattleboro rat. Ann. N.Y. Acad. Sci. 394, 1-802.

VALTIN, H.; SCHROEDER, H.A. (1964): Familial hypothalamic diabetes insipidus in rats (Brattleboro strain). Amer. J. Physiol. 206, 425-430.

VÖGELI, P.; STRANZINGER, G.; SCHNEEBELI, H.; HAGGER, C.; KÜNZI, N.; GERWIG, C. (1984): Relationship between the H and A-O blood types, phosphohexose isomerase and 6-phosphogluconate dehydrogenase red cell enzyme systems and halothane sensitivity, and economic traits in a superior and an inferior selection line of Swiss landrace pigs. J. Anim. Sci. 59, 1440-1449.

ZUCKER, L.M. (1965): Hereditary obesity in the rat associated with hyperlipemia. Ann. N.Y. Acad Sci. 131, 447-458.

Author's address:
V. Pliška, Institut für Nutztierwissenschaften, ETH Zürich, CH-8092 Zürich, Switzerland.

Gene and chromosome homologies in man and other mammals

G. F. STRANZINGER and R. HEDIGER

Introduction

Differences between the genomes and chromosomes in individual species are results of a long evolution. Similarities and homologies might have functional, or other, reasons. The human genome map has been the most extensively investigated in this respect, and approximately 3,000 genes have currently been assigned (HGM 9 1988). The genomes of farm animals have not been extensively studied until recently, but in the last few years valuable data have been collected and they can be used for several purposes (STRANZINGER 1987, WOMACK 1987), such as:
- marker-assisted selection for traits which have not normally been included in selection programs (i.e., resistance to various conditions, milk components, etc.);
- evaluation of autosomal recessive mutations causing effects on the gene products;
- molecular farming through the use of transgenic animals;
- monitoring long-term changes of gene frequencies due to environmental influences;
- investigation of evolutionary changes in the genome;
- establishing a basis for biomedical studies concerning parallels which exist between human and other mammalian genomes;
- investigation of aberrations in individual genes leading to genetic diseases, and variants of gene products, and comparing these to aberrations in different species.

Special emphasis will be put on the latter point.

Several programs have recently been launched around the world to map and to sequence the human genome (SINISCALCO 1987). Comparison with other mammals and non-mammalial species has become more important, since the information gained in this way may extend our understanding of the human genome. The basic genomic construction is similar in various species. Many functional and genetic homologies are already known, and mutational events with comparable functional consequences have attracted the attention of geneticists.

The topic of gene mapping

Human genetic diseases are caused by dominant, recessive and/or sex-linked mutations at single loci. More complex conditions originate from chromosomal aberrations and from genome mutations; these can also involve somatic mutations which are not normally inherited as traits subject to Mendel's rules of inheritance. The possibility that somatic mutations may lead, or contribute, to the cause of degenerative diseases of aging may be important enough for further investigations on genetic diseases. The burden of these diseases on the human population is known. Half of the spontaneous abortions (15 per cent of all recorded pregnancies) are caused by chromosomal imbalance. A high proportion of the human population suffers from cancer at some point in their lives. Genetic diseases and cancer are the primary reasons for genetic research in humans. At present over 3,000 human genetic diseases are known and, in about one fourth of the afflicted individuals, biochemical analysis can identify the defect (ORKIN 1982). Paradoxically low is the success rate of treatment, even when the defect is known. Phenylketonuria is one example in which avoidance of phenylalanine in the diet prevents the major symptom of the disease. Prenatal diagnosis is only applicable to families known to be at risk, since it is not currently possible to investigate the 400 diseases for which reliable tests are available at pregnancy. Research on human genetic diseases will lead to a better understanding of the diseases themselves and to a development of general principles to treat and cure these patients. Pharmaceutically important proteins and peptides, produced by recombinant techniques for replacement therapy may be a safe and relatively inexpensive tool for these purposes, but gene therapy of somatic cells remains the major goal. In order to achieve it, research will have to be conducted on animals. This research may indeed form the bridge between human and animal genetics. The interrelationships between developments in molecular genetics, cytogenetics and animal breeding have recently been outlined by FECHHEIMER (1986).

Animal breeding and production have very important roles in the world economy. As tools of agricultural production, animals are a decisive source of income for the farmer, and they play a vital role in the the efficient management of natural resources through the conversion of grasses and other foodstuffs into animal products.

The risk of excessive genetic manipulations should be prevented by the use of more refined methods of genetic engineering. Basic genetic research, including gene mapping and the investigations of the interactions between various gene loci within linkage and chromosomal constructions, will therefore become more important.

The construction of the genome

Many areas of genetics deal with the genetic information at the different levels: molecules, chromosomes, cells, gene products, or parameters of general traits measured in populations. Therefore, a broad range of possibilities is needed to explain and to understand the genome of an organism and the mutational events which may change the genetic structures and result in a modification of the genotype and phenotype. The phenomena of reproduction, internal metabolism, and sensitivity to environmental conditions express most of the functions of life within its genetic background. The normal make-up of a zygote and its consequent development depend on normal genes and chromosomes coming from both parents, in addition to a normal meiotic and mitotic history. The continuity of life depends on the functional capacity of the entire genome. Mutations at the level of genomic construction and function have affected evolution by substantially influencing selection and adaptation, and are still very important for the creation of new forms of genetic diversity (MAYR 1967). Changes of environmental conditions are not predictable, and thus biological systems face a variety of new conditions as well as the risk of self-extinction.

The chemical structure of genetic information

The chemical entity of inherited information and, particularly, the organizational structure of the polynucleotide chains is quite well understood. Of importance is, however, the gene expression and the enhancement of genetic information. Approximately 75 base pairs upstream from the gene locus lies the CAAT sequence ("CAT box"), where the transcription of mRNA is initiated and, approximately 30 base pairs upstream from the gene locus, the TATAAT sequence ("TATA box") is located. These two sets of sequences apparently help to direct the RNA polymerase to the proper initiation site so that it can transcribe the message of the gene to mRNA, from which it is then translated on the ribosomes to proteins. The CAT and TATA boxes form, together, the promotor region. The so-called "CAP site" is responsible for the addition of a 7-methylguanosine residue to the ends of the mRNA which may protect it against RNA-degrading enzymes. The start of the transcription site is an AUG sequence, and the end is marked by either UAG, UAA or UGA sequences. The gene to be transcribed contains *exons*, which are functional sequences coding for the gene product. The gene also contains *introns*, which are non-coding sequences that vary in size and number, and are interspersed between the coding sequences. In addition, there is a polyadenine tail (AUAAA) at the 5' end of the gene sequence. There may also be an enhancer sequence at some site inside or outside the gene sequence that enhances the transcriptional signal. All the above elements form the gene construct on different levels of translation and transcription and are necessary for the definition of genetic information and for the functional operation of the gene (KNIPPERS 1985). Crucial for this analysis was the discovery of restriction enzymes. There exist a large number of endonucleases that recognize specific sites within a DNA chain. These enzymes enable the base sequences to be analyzed and specific manipulations to be carried out. They are so specific that single functional units can be replaced, deleted, added or changed at will. The description "recombinant DNA technology" is derived from these possible precise manipulations (WATSON et al. 1985).

The microscopic structure of genetic information

Chromosomes consist of a large chain of nucleotides that are covered by histone and non-histone proteins. Structures can be seen only during mitosis and meiosis, using special preparation techniques. The size of a chromosome can be measured both in morphological units (in µm), and in chemical units (kilobases, kb and megabases, mb). Chromosomes are self-duplicating structures which are identical in number, form and organization for each species.

Chromosomes can be prepared and characterized by various culture and preparative methods, and from different cell and tissue samples. They may be stained differently by various dyes, which may indicate significant underlying functions. Basic histones take up Giemsa or Orcein stains, DNA is stained with Feulgen reagent, but no bands are visible unless additional steps are undertaken. Specific areas within the DNA sequence display different bands. For example, in A-T rich regions Quinicrine fluorescence produces identical bands (Q-bands) to Giemsa-trypsin treatment (G-bands), whereas treatment with 5-bromodeoxy-uridine (BrdU) which signify G-C rich regions. C-bands indicate heterochromatin which is concentrated around centromere regions. Telocentric chromosomes usually show a polymorphic pattern of the C-bands and are therefore very useful markers. Silver staining is specific for nucleolus organizer regions. Some of these techniques can be used in combination, which increases the amount of information gained. Chromosome banding techniques have proven to be particularly important for the development of gene mapping. First, they enabled a secure identification of each single chromosome in karyotypes, and second, they facilitated the standardization of chromosomes for international comparisons. For human chromosomes the ISCN Standard (1985) is accepted worldwide, and ever since the first Human Gene Mapping Workshop at the New Haven Conference in 1973, every second year a standardization workshop has taken place.

In farm animals, banded chromosomes were first discribed at the Reading Conference in 1976, and the results were published in the Reading Standard (1980). In addition to the standardization, use of this information enables the precise identification of individual mutational events on chromosomes. The size of a mammalian chromosome can vary considerably, but on average a metaphase chromosome is about 10 µm long and can contain any number of genes from 1,000 to 10,000. More specifically, it is estimated that there are about 3.0×10^9 bp in the haploid genome of a single cell, which translates to an average-sized chromosome of about 1×10^8 bp. If an average gene contains about 2,500 bp, then there are approximately 5,000 genes per chromosome. By applying the band staining to normally extended chromosomes, one can produce 5 to 10 (positive and negative) bands on an average-sized chromosome, resulting in 100 to 500 genes per band region. It is helpful to keep these relations in mind, while reading the literature or working with gene mapping techniques, since they are often utilized to describe the localization of a specific gene, or to estimate linkage and recombination frequencies.

Morphologic and genetic diversity of karyotypes

Human cytogenetic research is well advanced and a large amount of literature is available, as compared to animal cytogenetics. The usefulness of this achievement is apparent when examining the successful application of this discipline in many research areas. The standardization of the nomenclature for human chromosomes, based on their morphology, banding structure and mutational variation, is well documented, and accepted worldwide. The situation in the field of animal cytogenetic is, unfortunately, less satisfactory. Many scientists have published proposals and recommendations for the establishment of new standards for a particular species. However, some of these are based on misleading observations and thus are not useful. Nevertheless, the goal of establishing cytogenetic standards for animals remains important, and some of the reasons for this will be discussed here.

The elucidation of species diversity and evolutionary developments is of central interest. Identification of homologies of morphological structures of chromosomes, in addition to their genetic content, will permit extrapolation of the information gained to specific genetic and chromosomal aberrations. Chromosomal mutations arise *de novo* or are present at various frequencies in populations of different species. Polymorphisms in chromosomal structures can be used as markers but must be studied in detail before using them to analyze the inheritance of morphological traits. Artificial chromosomal mutations induced by mutagens should be evaluated so that they can be used for gene and chromosome mapping. Finally, in many genetic engineering experiments such as gene transfer, its integration, localization and gene expression should be determined in order to evaluate better the organization of the normal genome.

Table 1. Chromosome numbers (2n) of man and different species.

species	number of chromosoms (2n)
man (*Homo sapiens*)	46
cattle (*Bos taurus, Bos indicus*)	60
yak (*Bos grunniens*)	60
gaur (*Bos gaurus*)	58
gayal (*Bos [Bibos] frontalis*)	58
american bison (*Bison bison*)	60
european bison, wisent (*Bos bonasus*)	60
goat (*Capra hircus*)	60
sheep (*Ovis aries*)	54
congo buffalo (*Syncerus caffer nanus*)	54
african buffalo (*Syncerus caffer*)	52
murrah buffalo (*Bubalus bubalis*)	50
swamp buffalo (*Bubalus arnee*)	48
musk ox (*Ovibus moschattus*)	48
reindeer (*Rangifer tarandus*)	70
dromedar, one hump (*Camelus dromedarius*)	74
domestic camel, two humps (*Camelus ferus*)	74
guanako (*Lama guanicoe*)	74
vicunia (*Vicunia vicunia*)	74
horse (*Equus caballus*)	64
wild horse (*Equus caballus przewalskii*)	66
donkey (*Equus asinus*)	62
domestic pig (*Sus scrofa domestica*)	38
european wild pig (*Sus scrofa*)	36
rabbit (*Oryctolagus cuniculus*)	44
cat (*Felis cattus*)	38
dog (*Canis familiaris*)	78
wolf (*Canis lupus*)	78
chicken (*Gallus domesticus*)	78

The differences in karyotypes between man and farm animals are shown in Figs. 1 and 2. Chromosomal morphology (i.e., karyotype) and the number of chromosomes display characteristic differences. Table 1 summarizes the diploid number of chromosomes of several mammalian species, including man.

Methods for the localization of genes on chromosomes

Several methods for localizing genes onto chromosomes are known (RUDDLE 1981). Gene mapping began with the study of human families carrying visible anomalies which could be traced to a chromosomal mutation through pedigrees. In addition, it should be kept in mind that establishing a linkage (i.e., the genetic or morphologic distance between two loci is known) or synteny (i.e., the distance is not known) does not necessary include the mapping of the loci onto a specific chromosome. Therefore, the numbers of linkage or syntenic groups are not identical to their numbering after an assignment onto a chromosome. Chromosome markers (like polymorphic traits and mutated chromosomes) or, in the case of somatic cell hybrids, precisely identified chromosomes with their banding structure, are needed for the assignment.

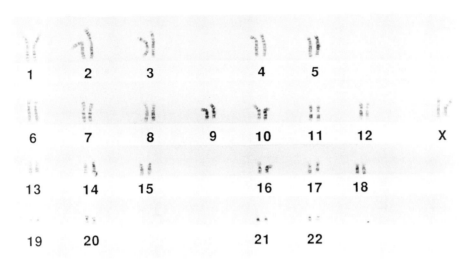

Fig. 1. Female human karyotype, GTG-banding. Courtesy of Prof. Dr. W. Schmid, University of Zurich.

Family investigations for recombination studies and assignments to chromosomes

Phenotypically visible syndromes without a major influence on the reproductive behaviour of afflicted individuals, spread within families, can often lead to a clarificaton of inheritance. In humans a survey of the chromosomes may result in the discovery of a common chromosomal mutation, and a systematic survey might reveal a linkage relationship between the syndrome and the mutation. However, not every phenotypically visible syndrome is the result of a mutation. Likewise, not every mutation causes a visible phenotypic variation. Specific statistical methods such as the lod score test (WALD 1947), which involves a sequential probability approach, are suitable to analyse the significance of the linkage situation. This method for family investigation has been widely used in human genetics since the very beginning of genome mapping. It can be also applied to farm animals, as was demonstrated by FRIES (1982) with the assignment of the G-blood group locus to chromosome 15 in swine. In human genetics, most material coming from hospitals and follow-up studies has facilitated additional investigations on linkage, and assignments to specific chromosomes. In farm animals the frequency of chromosomal mutations and related syndromes is quite low, partly because no animal breeder allows the propagation of animals with anomalies. Therefore, very limited use of this technique can be expected from practical field studies, and only on experimental farms will such an approach be possible. The artificial production of chromosomal mutations by irradiation of sperm can also be applied, as it has been used in the mapping approach for the G-blood group in swine (FRIES & STRANZINGER 1982). Exact mating programs are possible with animals and one should be able to use this technique to investigate special traits (blood group factors and other gene products).

Comparative mutation mapping

Very little is known about another potentially useful method called "comparative mutation mapping", which is an extension of the family studies mentioned above. This technique utilizes information about syndromes and chromosomal mutations in one species and compares it with the information available from another species. For example, the case of a naked female calf with tooth anomalies,

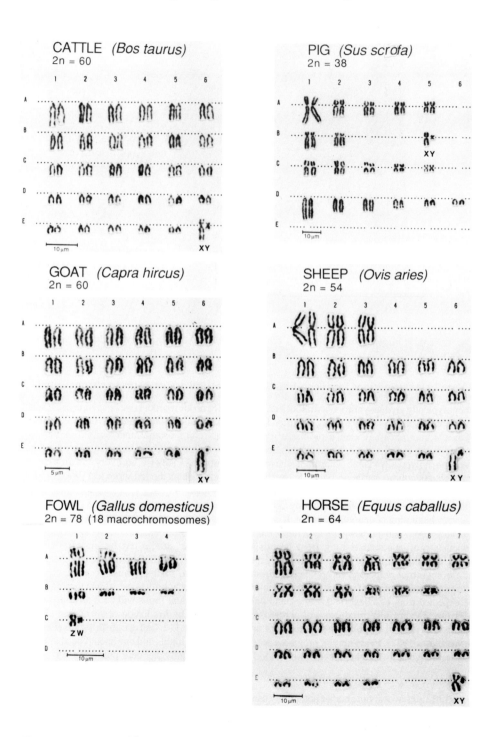

Fig. 2. Giemsa stained karyotypes of different farm animals.

an X-chromosome deletion and a reverse inactivation of the intact X-chromosome (BRAUN et al. 1988), has a counterpart in the human, where an X-chromosomal point mutation results also in hairlessness and tooth anomalies. Numerous other examples exist, and further use of this technique should be considered.

Somatic cell hybridization

With the help of different clones of somatic cell hybrids carrying identified chromosomes, and by analyzing species specific enzymes with polymorphic expression in these cell lines, a matrix structure of concordant and discordant situations for chromosomes and enzymes can be visualized. This permits the linkage of certain gene products and their gene loci to specific chromosomes (ROTH et al. 1986). Regional mapping to more defined areas of these chromosomes can be done by using induced chromosome mutations within each identified clone, with the expectation that in some cases the assigned chromosomes will carry new deletions. This, in turn, will change the concordant situation. Specific problems are generally caused in somatic cell hybrids by rearrangements of chromosomal segments in both species. Further techniques that can prove this situation in genomic DNA hybridization should be applied.

The *in situ* hybridization technique

The *in situ* hybridization technique makes use of the possibility of having gene probes of defined origin and base sequence hybridize with chromosomal DNA on the precise location where homologous sequences occur. Apparently, many gene probes of a single species can be used for hybridization in different species, as will be demonstrated in the following paragraphs. In order to visualize this region of hybridization, the probes are either radioactive or biotin labeling is employed. A precise identification of each metaphase chromosome within the many pre-identified metaphase spreads on several slides is necessary. The recording system of the hybridization signal should be made on sequentially stained and hybridized chromosomes which have been pre-identified and independently analyzed. The visible radioactive grains or regions stained by biotin are drawn into a histogram based on their precise position on the chromatid. The more pre-identified metaphases are analyzed, the more likely a significant peak will be established on a specific chromosome. This peak indicates the location of the homologous area (LIN et al. 1985).

Combined technique between molecular and cytogenetic methods

In combination with the above mentioned techniques and other molecular methods such as the Southern blot, a broad range of mapping possibilities has been established. With the help of restriction fragment length polymorphisms (RFLPs) and variable number tandem repeats (VNTRs), assignments of many genes to chromosomes in the human have been possible (HGM 9 1988).

Gene transfer

Gene transfer can be accomplished by introducing foreign genes into a cell or an embryo. The resulting organisms, which express the genes (i.e., produce the corresponding foreign protein), are called "transgenics" (CLARK et al. 1987). New genetic variants and polymorphisms can be created and studied *in vivo* under experimental conditions.

Gene and chromosome banding homologies in selected material

It is not surprising that for the phenotypic expression of certain mutations in different animal species, a common gene may be responsible. Comparative studies of mutations or color inheritance have long been the only indication for the genetic background. In recent years, however, it has become possible to extend these studies on the basis of molecular genetic information, and this has provided a new approach for gene mapping. For mutageneticut studies and extrapolation to the human, this information may prove very useful, and many experiments can then be substituted for by a general examination of a data base in the gene library. Selected examples of published experiments allow us to explain this in more detail.

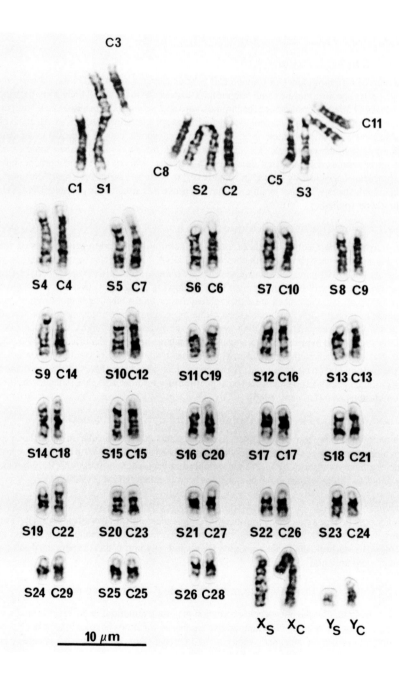

Fig. 3a. Comparison of karyotypes of cattle and sheep. C, cattle; S, sheep.

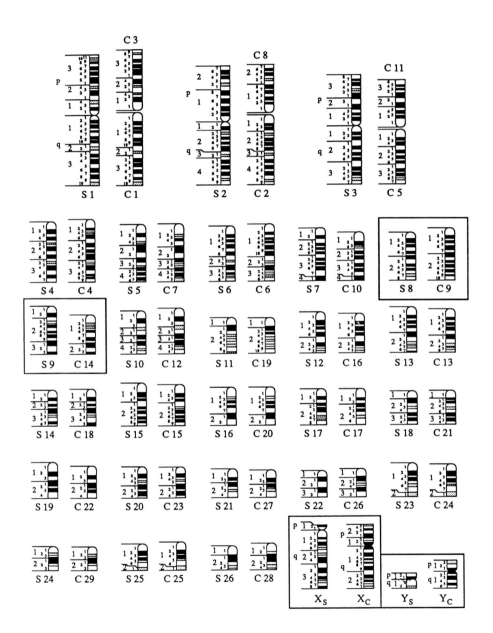

Fig. 3b. Comparison of idiograms of cattle and sheep. C, cattle; S, sheep.

Gene and chromosome banding homologies by *in situ* hybridization

The first assignment of the lymphocyte antigen (SLA) (GEFFROTIN et al. 1984) to the long arm of chromosome 7 in swine was accomplished by using a human gene probe for the antigens of the major histocompatibility complex (MHC). RABIN and coworkers (1985) used a porcine class 1 gene probe and mapped the SLA locus again to chromosome No. 7. However, contrary to the paper mentioned above, the locus was mapped to the short arm of the submetacentric chromosome. This latter assignment was confirmed by ECHARD and coworkers in 1986 by using a gene probe different from that used by the first two groups. These results indicate two things. First, in the initial assignment the identification of the chromosome region was probably limited by the number of metaphases examined, by the skill of the investigators, and by the statistical approach used to construct the histogram. Second, it is clear that different gene probes are able to hybridize to the same specific chromosomal region.

We extended this study with the SLA class 1 probe to the horse (ANSARI et al. 1988), cattle and sheep (HEDIGER 1988). In all three species, the hybridization peak was significant for one specific chromosome: chromosome No. 20 for the horse; chromosome No. 23 for cattle; and chromosome No. 20 for sheep.

Another interesting result was obtained by this approach. Cattle chromosome No. 23 and sheep chromosome No. 20 (after the Reading Standard, with modifications proposed by HEDIGER 1988) were shown to be identical in G-banding appearance (Fig. 3). It was thus documented for the first time that within the bovidae chromosome, banding homologies correspond to gene homologies. In fact, only two significant banding differences exist between the entire karyotypes of cattle and sheep. Further evidence for these homologies are shown in Fig. 4 for the keratin genes, A and B. In this case, it is possible even to compare the metacentric sheep chromosome No. 3 with the telocentric cattle chromosome No. 5 in such a way that the G-banding patterns of sheep chromosome No. 3 resemble the fusion of cattle chromosomes Nos 5 and 11. Therefore, it is not surprising that cattle chromosome No. 5 and the long arm of sheep chromosome No. 3 have similar banding patterns. Furthermore, according to the homologous hybridization of the keratin B gene, an evolutionarily conserved region in both banding and gene location still exists. This is in accordance with the keratin A assignment to cattle chromosome No. 19 and sheep chromosome No. 11 since, again, these two chromosomes show complete banding homology.

Extension of linkage homologies to several species including man

Using the information given at HGM 9 (1988) on the latest results of gene assignments in man and other species, and taking into account the published results of FRIES et al. (1988) and FRIES (1989) on the linkage groups of the genes for hemoglobin beta (HBB), parathyroid hormone (PTH), and follicle stimulating hormone (FSH) in cattle, one is able to show the following in Fig. 5: the linkage group of 10 loci on the short arm (p) of human chromosome No. 11 consists of the loci for HBB, PTH and FSH in a linear order surrounded by other gene loci which have not yet been assigned in cattle. The gene loci for HBB, PTH and FSH have been assigned to cattle chromosome No. 15. For comparison, it can be shown that these gene loci are already split between two chromosomes in the mouse, as HBB and PTH are on chromosome No. 7 and as FSH is on chromosome No. 2. In the cat, since not even FSH has been assigned, one should expect this locus to be on chromosome D1 because lactate dehydrogenase A (LDHA) and acid phosphatase 2 (ACP2) in man surround FSH; therefore, the likelihood of their being contained in this linkage group is very high. The situation also occurs in the chimpanzee chromosome No. 9. Use of information from the extensively mapped human genome will enable further mapping of the genomes of other mammalian species. The availability of this information is guaranteed by the quality of the different human gene mapping libraries and computer services as, for example, by the New Haven Human Gene Mapping Library of the Howard Hughes Medical Institute at Yale University. Unfortunately, the ever increasing amount of data on farm animal gene mapping is not included in this library; this is a major problem which requires a prompt solution.

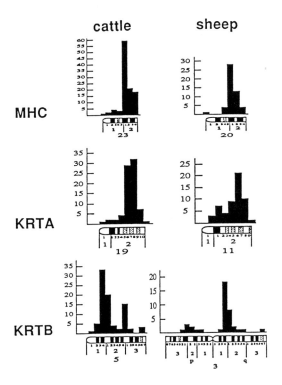

Fig. 4. In situ hybridization with cattle and sheep chromosomes with two keratin gene probes (KRTA and KRTB) and with the major histocompatibility complex (MHC) of swine lymphocyte antigene (SLA) class I gene probe.

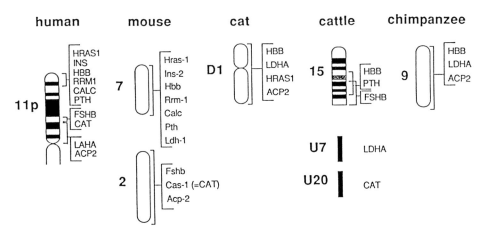

Fig. 5. Comparison of linkage groups containing the genes of hemoglobin beta (HBB), parathyroid hormone (PTH), follicle stimulating hormone (FSH) and other genes, in different species including man. The profiles demonstrate evolutionary conservation of larger areas in chromosomes.

Summary

Gene mapping and chromosome analysis are important topics in human and animal genetics. Several national and international programs invest academic and financial resources into this field, and the use of the information made available should be guaranteed. The application of such information is conceivable in several disciplines. A description of the genome may be useful for understanding homologies between the genes and chromosomes of different mammalian species. Different approaches to localize genes are described and several examples are given of gene and chromosome banding homologies between man, farm, laboratory and wild animals. The coordination of efforts in gene mapping is necessary, and may help to understand the construction and function of the mammalian genome.

References

ANSARI, H.A.; HEDIGER, R.; FRIES, R.; STRANZINGER, G. (1988): Chromosomal localisation of the major histocompatibility complex of the horse (ELA) by *in situ* hybridisation. Immunogenetics 28, 362-364.
BRAUN, U.; ANSARI, H.A.; HEDIGER, R.; SÜSS, U.; EHRENSPERGER, F. (1988): Hypotrichose und Oligodontie, verbunden mit einer Xq-Deletion, bei einem Kalb der Schweizerischen Fleckviehrasse. Tierärztl. Prax. 16, 39-44.
CLARK, A.J.; SIMONS, P.; WILMUT, I.; LATHE, R. (1987): Pharmaceuticals from transgeneic livestock. Trends Biotechnol. 5, 20-24.
DAVIES, K.E: (1988): Genome Analysis: A Practical Approach. IRL Press, Oxford, Washington DC.
ECHARD, G.; YERLE, M.; GELLIN, J.; DALENS, M.; GILLOIS, M. (1986): Assignment of the major histocompatibility complex to the p1.4 - q1.2 region of chromosome 7 in the pig (*Sus scrofa domestica L.*) by *in situ* hybridisation. Cytogenet. Cell Genet. 41, 126-128.
FECHHEIMER, N.S. (1986): Interrelationships between recent developments in molecular genetics and cytogenetics and animal breeding. J. Dairy Sci. 69, 1743-1751.
FRIES, R. (1982): Natürliche und induzierte Markerchromosomen beim Schwein und ihre Verwendung für die Genkartierung. Thesis No. 7160, ETH Zürich.
FRIES, R. (1989): The gene for the beta-subunit of the follicle-stimulating hormone maps to bovine chromosome 15. J. Heredity (submitted).
FRIES, R.; HEDIGER, R.; STRANZINGER, G. (1988): The loci for parathyroid hormone and beta-globin are closely linked and map to chromosome 15 in cattle. Genomics 3, 302-307.
FRIES, R.; STRANZINGER, G. (1982): Chromosomal mutations in pigs derived from X-irradiated semen. Cytogen. Cell Genet. 234, 55-66.
GEFFROTIN, C.; POPESCU, C.P.; CRIBIU, E.P.; BOSCHER, J.; RENARD, C.; CHARDON, P.; VAIMAN, M. (1984): Assignment of MHC in swine to chromosome 7 by *in situ* hybridisation and serological typing. Ann. Génét. 27, 213-219.
HEDIGER, R. (1988): Die *in situ* Hybridisierung zur Genkartierung beim Rind und Schaf. Thesis No. 8725, ETH Zürich.
HGM 9 (HUMAN GENE MAPPING, WORKSHOP 9) 1988: Paris Conference (1987). Cytogenet. Cell Genet. 46, 1-762.
ISCN STANDARD (1985): An international system for human cytogenetic nomenclature. Report of the standing committee on human cytogenetic nomenclature. Cytogenet. Cell Genet. Karger. Basel, New York.
KNIPPERS, R. (1985): Molekulare Genetik. Georg Thieme Verlag. Stuttgart, New York.
LIN, C.C.; DRAPER, P.N.; DE BRAELELEER, M. (1985): High-resolution chromosomal localisation of the beta-gene of the human beta-globin gene complex by *in situ* hybridisation. Cytogenet. Cell Genet. 39, 269-274.
MAYR, E. (1967): Artbegriff und Evolution. Verlag Paul Parey. Hamburg und Berlin.
ORKIN, S.H. (1982): Genetic diagnosis of the fetus. Nature 296, 202-203.
RABIN, M.; FRIES, R.; SINGER, D.; RUDDLE, F.H. (1985): Assignment of the porcine major histocompatibility complex to chromosome 7 by *in situ* hybridisation. Cytogenet. Cell Genet. 39, 206-209.
READING STANDARD (1980): Proceedings of the first international conference for the standardisation of banded karyotypes of domestic animals in Reading 1976. (Editors: FORD, C.E.; POLLOCK, D.L.; GUSTAVSSON, I.). Hereditas 92, 145-162.
ROTH, H.R.; DOLF, G.; STRANZINGER,G. (1987): Gene assignment by means of somatic cell hybrids: a new method for the analysis of data. Theor. Appl. Genet. 74, 42-48.
RUDDLE, F.H. (1981): A new era in mammalian gene mapping: somatic cell genetics and recombinant DNA methodologies. Nature 294, 115-120.
SINISCALCO, M. (1987): On the strategies and priorities for sequencing the human genome: a personal view. Trends Genet. 3, 182-184.

STRANZINGER, G. F. (1987): Gene mapping and gene homologies in farm animals: techniques and present status of gene maps. Anim. Genet. **18**, 111-116.
WOMACK, J.E. (1987): Genetic engineering in agriculture: Animal genetics and development. Trends Genet. **3**, 65-68.
WATSON, J.D.; TOOZE, J.; KURTZ, D.T. (1983): Recombinant DNA. A short course. Scientific American Books, Inc., NY.

Authors' address:
G. Stranzinger and R. Hediger, Institut für Nutztierwissenschaften, ETH Zürich, CH-8092 Zürich, Switzerland.

Physiology and genetics

Farm animals as models of cholesterol metabolism

A. C. BEYNEN

Introduction

The concentration of cholesterol in the blood of man is an important risk indicator for coronary heart disease. The higher this concentration the higher is the risk for cardiovascular disease. This notion has been derived from various types of studies such as epidemiological and case-control studies in man, and also from controlled studies with animal models. Serum cholesterol may be a causative factor in the development of coronary heart disease. Lowering of serum cholesterol levels has been shown to decrease the incidence of cardiovascular mortality (Lipid Research Clinics Program 1984).

As cardiovascular disease is the leading cause of death in western countries, there is a great deal of interest in the regulation of cholesterol metabolism. Knowledge of the factors involved in the control of serum cholesterol concentrations might contribute to the optimal prevention and therapy of cardiovascular disease in man. Animal models are used in order to gain further insight into the mechanisms controlling human serum cholesterol concentrations. The value of animal models is therefore critically dependent upon their degree of similarity with man. Thus, progress in cholesterol metabolism research is closely associated with the selection of appropriate animal models. This paper compares farm animals and common laboratory animals, as possible models of cholesterol metabolism.

Cholesterol in serum

Cholesterol in serum is carried by lipoprotein particles. On the basis of their density and electrophoretic properties, three major classes of lipoproteins have been separated: very low density lipoproteins (VLDL), low density lipoproteins (LDL) and high density lipoproteins (HDL). The limits of density for human lipoproteins are: VLDL < 1.006 g/ml, LDL 1.019 - 1.063 g/ml, and HDL 1.063 - 1.210 g/ml. The proportion of total cholesterol in serum carried by these lipoproteins varies among animal species (TERPSTRA et al. 1982).

By density gradient ultracentrifugation of serum which was pre-stained with Sudan black, lipoproteins can be separated and visualized. Fig. 1 shows the serum lipoprotein profiles of various animal species: VLDL at the top of the ultracentrifugation tube, LDL and HDL further down, and a residual stain is at the bottom of the tube. A few striking similarities, and dissimilarities, can be noted. The profiles of humans, pigs and chickens are similar in that they all have a band in the LDL density range. Other animals show a very pronounced HDL band. The lipoprotein profiles of the cow and rat are almost identical. On the other hand, the profiles of the two ruminants (cows and goats) are different.

Cholesterol metabolism

As a structural component of most cell membranes, cholesterol is an essential compound for all animals, including man. In addition, cholesterol is used as the substrate for the synthesis of steroid hormones, vitamin D and bile acids. In order to fulfil the cholesterol demands of various cells, a continuous supply of cholesterol is required.

During growth, cholesterol accumulates with new tissues, and hence the body pool of cholesterol increases. However, at steady-state, the amount of cholesterol in the body is determined by the balance of input and output. Cholesterol enters the body pool from two sources: absorption from the diet and synthesis within the various organs of the body. Cholesterol leaves the body by various routes. Hepatic cholesterol can be transferred as such to the bile fluid and then to the gastrointestinal tract. Alternatively, cholesterol can be converted first into bile acids. Cholesterol and bile acids that are not subjected to re-absorption eventually leave the body in feces. Cholesterol is also lost from the body through the

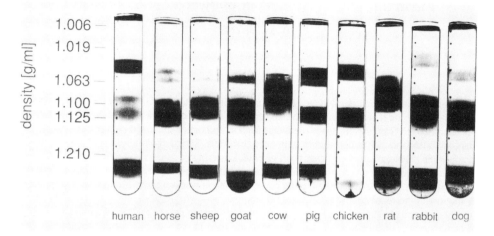

Fig. 1. Serum lipoprotein patterns observed after density gradient ultracentrifugation of pre-stained serum (TERPSTRA et al. 1982).

sloughing of both intestinal mucosa and skin. The cholesterol converted into steroid hormones is excreted from the body in urine. Quantitatively, fecal excretion of cholesterol, its bacterial metabolites and bile acids is the most important excretion mechanism. Thus, at the steady-state, cholesterol synthesis in the body can be assessed as the sum of fecal excretion of neutral steroids and bile acids *minus* cholesterol intake. The concentration of serum cholesterol is determined by the same factors as the body pool of cholesterol. In addition, serum cholesterol can be influenced by changes in the distribution of cholesterol between serum and tissues, and by the capacity of tissues to store cholesterol.

Cholesterol homeostasis

To maintain the balance of cholesterol throughout the body, a series of regulatory and compensatory mechanisms have been evolved in different animal species and man. The salient features of these mechanisms can be illustrated by describing the cascade of events which occur in response to increased intakes of cholesterol. The efficiency of cholesterol absorption probably remains constant after cholesterol feeding and thus, absorption might not be considered a regulatory mechanism. However, the efficiency of cholesterol absorption can differ markedly between animal species as well as among the strains of one species (Fig. 2), and it appears to depend upon the composition of the diet.

The maximum set of compensatory mechanisms triggered by cholesterol loading consists of the following processes. An increased intake of cholesterol will cause an increased influx of cholesterol into the liver, delivered to this organ by the so-called chylomicron-remnants. This increased influx tends to enlarge liver cholesterol pools. The liver responds immediately by inhibition of *de novo* cholesterol synthesis so that no extra cholesterol is added to the liver cholesterol pools. The liver also possesses mechanisms for removing an excess of cholesterol. Cholesterol can be channeled into bile fluid either as such, or after conversion into bile acids. In addition, cholesterol can be secreted into serum as a component of lipoprotein particles. On the other hand, cholesterol uptake in the form of lipoprotein-cholesterol can be depressed by down regulation of the activity of the receptors (apo B, E receptor) on liver membranes. The balance between cholesterol synthesis plus uptake on the one hand, and lipoprotein cholesterol output by the liver on the other, will be reflected in the fecal excretion of neutral steroids and bile acids. In essence, cholesterol can only leave the body through the liver. The compensatory mechanisms act cooperatively so that, in the long run, a new steady-state will be reached. At this new steady-state the concentration of cholesterol in serum may be somewhat

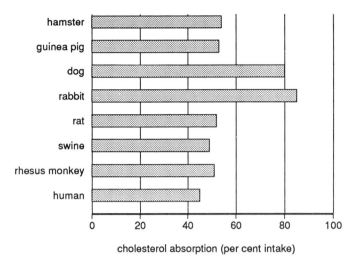

Fig. 2. *Approximate efficiency of cholesterol absorption in various species. From BEYNEN (1988).*

increased, the degree of this increase being dependent upon the capacity of the compensatory mechanisms.

In the following paragraphs various animal species will be compared with regard to their response of cholesterol synthesis and fecal bile acid excretion to cholesterol feeding. Evidence will be presented that cholesterol metabolism and its compensatory mechanisms differ among animal species in both qualitative and quantitative terms.

Cholesterol synthesis

Whole-body cholesterol synthesis can be derived from sterol-balance data; that is, calculated as the difference between cholesterol excretion in feces (sum of neutral steroids and bile acids) and cholesterol intake. A prerequisite is that the animal is in a steady-state so that the input equals to the output. Due to the accumulation of cholesterol in the body such a situation is not reached in many short-term cholesterol feeding experiments. In growing animals there is a formation of new tissues which contain membrane cholesterol, in addition to cholesterol storage. This leads to the calculation

Table 1. *Approximate rates of whole-body cholesterol synthesis (in mg/day/kg body weight) in different animal species fed low or high cholesterol diets. (BEYNEN 1988).*

	cholesterol ingested		cholesterol synthesis	
	low	high	low	high
man	2	10	10	7
rhesus monkey	0	63	18	11
swine	0	41	57	9
rat	39	255	62	27
dog	1	115	13	0
guinea pig	2	30	24	19
rabbit	0		25	

of falsely low or even negative rates of cholesterol synthesis. Thus, cholesterol accumulation should be corrected for.

Table 1 compares the rates of whole-body cholesterol synthesis determined on low and high-cholesterol diets in different animal species, including man. It appears that there are no qualitative differences between species. Cholesterol feeding invariably reduces the whole-body cholesterol synthesis. No definitive conclusions can be drawn as to possible quantitative differences in the response to dietary cholesterol, since cholesterol intakes differed per species.

Table 2. Fecal excretion of bile acids (in mg/day/kg body weight) on low and high cholesterol diets in different species. (BEYNEN 1988, BEYNEN et al. 1983).

	cholesterol ingested		cholesterol synthesis	
	low	high	low	high
man	2	10	3	3
rhesus monkey	0	63	4	9
swine	0	41	28	12
veal calf	8	154	1	1
rat	7	192	28	121
rabbit	3	100	17	19
dog	1	115	5	48

Bile acid excretion

Cholesterol feeding most likely induces an increase in the fecal output of neutral steroids in all animal species. Most of this increase probably represents non-absorbed cholesterol rather than hepatic cholesterol excreted through bile. Stimulation of bile acid excretion upon feeding cholesterol does not occur in all species (Table 2). In man, swine, pre-ruminant calf, rabbit, and perhaps the rhesus monkey, fecal bile acid output is not enhanced. In contrast, the rat and dog show a dramatic increase of bile acid excretion. Thus, there are qualitative differences of bile acid excretion between animal species in the response to cholesterol loading, unlike the case of neutral steroid excretion. Details are beyond the scope of this communication, but it should be stressed that the relative proportions of individual neutral steroids and bile acids can differ markedly between species.

Theoretically, it is possible that the inter-species variation in the ability to step up bile acid excretion after cholesterol loading is a quantitative feature rather than a qualitative one. It might be suggested that in certain species an increase in bile acid excretion only occurs when inhibition of cholesterol synthesis cannot compensate for the increase in the amount of cholesterol absorbed. This might mean that the rate of whole-body cholesterol synthesis should be considered for the correct choice of an animal model. Alternatively, the amount of dietary cholesterol should be tailored to the rate of cholesterol synthesis. If the increase in the amount of cholesterol absorbed is much higher than the basal rate of cholesterol synthesis, probably irrespective of whether bile acid synthesis is activated or not, cholesterol will be accumulated in the body, especially in the liver. This may cause liver damage and diminished liver function (BEYNEN et al. 1986a), which in turn could lead to a biased interpretation of experimental results.

Serum cholesterol response to dietary cholesterol: inter-species differences

The interaction between compensatory mechanisms and cholesterol feeding determines the magnitude of change seen in serum cholesterol. In species with relatively limited capacity to absorb cholesterol, e.g. man, a small increase in dietary cholesterol will be compensated for by inhibition of cholesterol synthesis, and serum cholesterol will rise only slightly, if at all. If the amount of absorption exceeds the quantity synthesized by the body, and if conversion into bile acids cannot increase, the pool of serum cholesterol will rise dramatically. This situation occurs in cholesterol feeding studies with rabbits. If the animal species (such as the rat) possesses both the ability to depress high basal rates of cholesterol synthesis, and the ability to enhance bile acid production, serum cholesterol does not necessarily increase. Thus, after cholesterol consumption, serum cholesterol concentration is controlled by the efficiency of absorption, inhibition of cholesterol synthesis and the ability to activate bile acid synthesis.

Serum cholesterol response to dietary cholesterol: intra-species differences

The addition of cholesterol to the diet of random-bred rabbits elicits a rise of serum cholesterol but many investigators have noted that there are marked inter-individual differences in the intensity of the response. This indicates that certain animals are hypo-, and others are hyperresponsive to dietary cholesterol, as illustrated in Fig. 3. Extreme differences in the response of serum cholesterol to diet can be found among inbred strains of rabbits. Table 3 shows the levels of serum cholesterol in male rabbits of six inbred strains. The rabbits were sampled while they were on a commercial rabbit chow, and also after 21 days of receiving the same diet to which 0.5% (w/w) cholesterol had been added. The animals with the most extreme response showed almost a 5-fold higher increase in serum cholesterol than the strain with the lowest response.

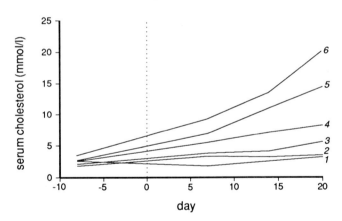

Fig. 3. Effect of cholesterol feeding (as from day 0) on serum cholesterol concentrations in individual (1-6) random bred New Zealand White rabbits (BEYNEN et al. 1986b).

Recently some insight into the mechanisms underlying the difference in cholesterolemic response between the strains displaying the most extreme reactions (IIIVO/J and AX/J) has been obtained (BEYNEN et al. 1989). The hyporesponsive strain was found to slightly enhance its excretion of bile acids after cholesterol feeding, whereas the hyperresponders did not. Furthermore, the hyporesponsive rabbits absorbed dietary cholesterol significantly less efficient than did their hyperresponsive counterparts.

Studies with inbred strains of rats (VAN ZUTPHEN & DEN BIEMAN 1981), mice (BEYNEN et al. 1987) and pigeons (WAGNER & CLARKSON 1974) have also demonstrated marked strain differences in the response of plasma cholesterol to cholesterol-rich diets. These studies indicate that differences in responsiveness have a genetic basis.

Table 3. *Effect of dietary cholesterol on plasma cholesterol levels in inbred strains of rabbits.*

strain	plasma cholesterol (mmol/l)[a]	
	day 0	day 21
IIIVO/J	0.7 ± 0.1	8.0 ± 1.9
WH/J	0.3 ± 0.0	8.0 ± 0.8
X/J	0.9 ± 0.3	17.3 ± 4.2
ACEP/J	0.7 ± 0.0	18.9 ± 3.5
OS/J	0.8 ± 0.1	29.4 ± 4.8
AX/J	0.6 ± 0.1	33.6 ± 3.1

[a]*Results are expressed as arithmetic means ± standard errors (5 male animals per strain). Until day zero of the experiment all rabbits received a commercial cholesterol-free diet; then, 0.5% (w/w) cholesterol was added to the diet. Data taken from* VAN ZUTPHEN & FOX *(1977).*

Human hypo- and hyperresponders

In the numerous studies which have dealt with the effect of dietary cholesterol on serum cholesterol levels in humans, a striking variability in individual response has been generally found. Certain individuals show negligible changes in the concentration of serum cholesterol (hyporesponders), whereas others develop elevated concentrations (hyperresponders). The concept of human hyper- and hyporesponders has become firmly entrenched in the literature. However, in almost all studies, the dietary challenge has been only given once and, thus, the reproducibility of the individual cholesterolemic response is not known.

We have carried out three controlled dietary trials with the same subjects, to address the question as to whether individuals do exist with a consistently high or low serum cholesterol response to dietary cholesterol (KATAN et al. 1986). In each trial, the volunteers successively consumed a low- and a high-cholesterol diet, the cholesterol component of the diets (provided by egg yolk) being the only variable. Standardized regression coefficients for individual responses in one experiment as a function of the response in a previous experiment ranged from 0.34 to 0.53 (n = 32, p < 0.05).

It thus appears that at least a part of the cholesterolemic response to dietary cholesterol in man is individually determined, although the range of responsiveness is much smaller in man than in laboratory animals. It is also clear that one will always find subjects who appear hyperresponsive in one experiment and hyporesponsive in another. This is caused by the diet-independent within-person variability of serum cholesterol. Nevertheless, from these repeated experiments with the same subjects, it can be concluded that human hypo- and hyperresponders to dietary cholesterol do exist.

Heritability of differences in responsiveness

The question could be raised as to what extent differences in response between random-bred animals are genetically determined. The variance of any biological trait is, of course, always composed of genetic and environmental contributions. If the environment is made more uniform, then the apparent heritability of the trait increases, and if the environmental conditions fluctuate a lot from animal to animal then genetic determinants are obscured. Thus, heritabilities estimated in different studies can usually not be compared. Still, it is legitimate to ask what proportion of the variance in response observed in animal experiments is due to genetic differences, and what proportion is due to metabolic idiosyncrasies that are fixed at another stage, for example, early nutrition. The extent of the genetic influence on variability in serum cholesterol response to dietary cholesterol has been studied in several animal species. In wild-type squirrel monkeys, CLARKSON et al. (1971) found that about 65% of the variability in serum cholesterol concentration, after cholesterol feeding was attributable to genetic factors.

ROBERTS et al. (1974) studied a large population of random-bred rabbits, and obtained a wide range of plasma concentrations both on a commercial diet and on a diet containing 0.28% cholesterol. A breeding trial with selected hyper- and hyporesponsive animals showed that the response of plasma cholesterol to dietary cholesterol was inherited. The heritability, estimated from the regression of mean progeny response on midparent response, was found to be 50%.

VAN ZUTPHEN and DEN BIEMAN (1983) studied the inheritance of responsiveness in male rats obtained by crossing a hyperresponsive inbred strain with a hyporesponsive strain (SD/Cpb and SHR/Cpb). A diet containing 2% cholesterol and 0.5% cholate was used. The F1 hybrids had responses about halfway between those of the parental strains. Animals of the F2 hybrid had responses scattered over the entire scale. Comparison of the variances of the response of genetically uniform groups (parental strains and F1 hybrid), and segregating groups (backcrosses and F2 hybrid), revealed that under these specific conditions more than 80% of the observed variation could be attributed to additive genetic factors, and that two major genes were involved in the control of the serum cholesterol response. The studies with monkeys and rabbits described above demonstrated a weaker genetic influence on sensitivity to a hypercholesterolemic diet. However, as pointed out at the beginning of this paragraph, heritability cannot be measured in absolute units because it depends upon the specific environment in each experiment. Still, it can be concluded that a major part of the variability of response observed in laboratory animals under standard experimental conditions is due to genetic differences.

Response of serum cholesterol to dietary fatty acids

The fatty acid composition of the diet is a powerful determinant of serum cholesterol concentrations. Table 4 shows that rabbits fed cholesterol-free, semipurified diets responded to dietary fats in a similar way as humans do. Studies with rabbits may thus contribute to a description of the mechanisms

Table 4. Effect of various dietary fats on serum cholesterol in rabbit and man[a].

dietary fat	rabbit[b]	human[c]
soybean oil	0.00	0.00
corn oil	-0.21	0.00
cottonseed oil	+0.88	+0.20
olive oil	+2.12	+0.32
beef tallow	+2.82	+0.61
butter fat	+4.14	+0.77
coconut fat	+7.64	+1.09

[a]Rabbit data are based on studies reported by CARROLL (1971). The diets of the rabbits consisted of (g/100 g): casein, 30; dextrose, 44; fat, 15; salt mixture, 5; celluflour, 5; vitamin mixture, 1. The experiment lasted two weeks. The rabbits weighing about 1.2 kg consumed about 10 g of fat per day. Human data are theoretical effects of various fats on serum cholesterol in subjects consuming 100 g of fat in a diet providing 2250 kcal and 400 mg cholesterol per day, as predicted by the formula of KEYS et al. (1965).

[b]Change in serum cholesterol upon substitution of an indicated fat for soybean oil (mmol/l).

[c]Increase in serum cholesterol concentration (mmol/l) upon substitution of 50 g of a particular fat for 50 g of soybean oil.

underlying the differential cholesterolemic effects of dietary fats in humans. Rats fed a cholesterol-free diet may be an unsuitable model because, under such conditions, dietary polyunsaturated fatty acids in the form of corn oil cause higher serum cholesterol levels than saturated fatty acids given as coconut fat (BEYNEN & KATAN 1984). Veal calves also display increased concentrations of serum cholesterol after the feeding of polyunsaturated fatty acids when compared with saturated fatty acids (JACOBSON et al. 1974).

Response of serum cholesterol to dietary proteins

Dietary casein produces higher levels of serum cholesterol in rabbits than does soy protein. Table 5 illustrates that rabbits may be most sensitive in this respect. Guinea pigs, rats, swine, hamsters and

Table 5. Species-dependent increase in total serum cholesterol following diets containing soy protein with or without added cholesterol compared to a casein diet. (After WEST & BEYNEN 1986).

	average increase in serum cholesterol (mmol/l)	
	low cholesterol diet	high cholesterol diet
rabbit	5.9	14.5
guinea pig	0.5	3.3
swine	0.0	2.7
rat	0.0	2.3
hamster	0.4	2.3
monkey	0.1	1.3
chicken	-0.2	0.2
calf	-0.2	
mouse	-0.1	-0.2

monkeys also respond to the type of dietary protein, but not to the same extent as do rabbits, and only if the diet contains cholesterol. Chickens, calves and mice appear to be rather insensitive. It should be emphasized that Table 5 gives only a general impression of the species-dependent sensitivity to dietary proteins. Other components of the diet, such as the type of fat and fiber, and characteristics of the experimental animals, such as age, sex and strain may also determine the susceptibility to proteins in the diet. It is clear from Table 5 that the differential effect of casein and soy protein becomes more pronounced in diets enriched with cholesterol.

It could be argued that substances other than protein in the test diets were not the same in the studies with various animal species, including man. These dietary components may have overridden a possible protein effect in those species showing no differential response. This reasoning is supported by the observation that in rabbits an increase in the proportion of essential fatty acids, at the expense of saturated fatty acids in the diet, causes complete disappearance of the casein-induced hypercholesterolemia (BEYNEN & WEST 1981). Thus, only comparative studies with the same diet may give conclusive results as to species-dependent effects.

VAN RAAIJ et al. (1981) and VAN DER MEER et al. (1985) have reported the results of experiments in which rabbits were fed on the same diets as either humans or rats, respectively. In contrast to the human subjects and rats, the rabbits displayed a hypercholesterolemic response to casein, when compared to soy protein. This indicates that there are indeed true differences between species with regard to the response of serum cholesterol to dietary casein. The rabbit appears to be extremely sensitive. This was also borne out by a study in which we fed rabbits on the milk substitutes used for pre-ruminant calves. The milk substitute containing skimmed-milk powder significantly raised serum cholesterol in rabbits more than that with the substitute containing soy protein. In veal calves fed the same diets there was no such differential effect of the two diets (BEYNEN et al. 1983b).

Epilogue

Selection of the appropriate animal model depends upon the particular question under study, but knowledge of qualitative and quantitative differences in cholesterol metabolism among species is required. It is important that the method of challenging cholesterol metabolism, e.g. by cholesterol feeding, is tailored to the characteristics of cholesterol metabolism in the animal model under study. If compensatory mechanisms are overwhelmed and cannot handle the challenge, cholesterol may accumulate and lead to a biased interpretation of the results.

References

BEYNEN, A.C. (1988): Animal models for cholesterol metabolism studies. In: New Developments in Biosciences: Their Implications for Laboratory Animal Science (Editors: BEYNEN, A.C.; SOLLEVELD, H.A.). Martinus Nijhoff Publishers, Dordrecht, Pp. 279-288.
BEYNEN, A.C.; DANSE, L.H.J.C.; VAN LEEUWEN, F.X.R.; SPEIJERS, G.J.A. (1986a): Cholesterol metabolism and liver pathology in inbred strains of rats fed a high-cholesterol, high-cholate diet. Nutr. Rep. Int. **34**, 1079-1087.
BEYNEN, A.C.; KATAN, M.B. (1984): Relation between the response of serum cholesterol to dietary cholesterol and to the type of dietary fat, protein and fiber in hypo- and hyperresponsive rats. Nutr. Rep. Int. **29**, 1369-1381.
BEYNEN, A.C.; KATAN, M.B.; VAN ZUTPHEN, L.F.M. (1986b): Hypo- and hyperresponders to dietary cholesterol. In: Nutritional Effects on Cholesterol Metabolism (Editor: BEYNEN, A.C.). Transmondial, Voorthuizen. Pp. 99-109.
BEYNEN, A.C.; LEMMENS, A.G.; DE BRUIJNE, J.J.; RONAI, A.; WASSMER, B.; VON DEIMLING, O.; KATAN, M.B.; VAN ZUTPHEN, L.F.M. (1987): Esterases in inbred strains of mice with differential cholesterolemic responses to a high-cholesterol diet. Atherosclerosis **63**, 239-249.
BEYNEN, A.C.; MEIJER, G.W.; LEMMENS, A.G.; GLATZ, J.F.C.; VERSLUIS, A.; KATAN, M.B.; VAN ZUTPHEN, L.F.M. (1989): Sterol balance and cholesterol absorption in inbred strains of rabbits hypo- or hyperresponsive to dietary cholesterol. Atherosclerosis **77**, 151-157.
BEYNEN, A.C.; VAN GILS, L.G.M.; GOLDSMID, A.W.; SCHOUTEN, F.J.M.; HECTORS, M.P.C. (1983a): Digestion of the diet and fecal excretion of neutral and acidic steroids by veal calves fed a milk replacer supplemented with cholesterol. Nutr. Rep. Int. **28**, 825-833.
BEYNEN, A.C.; VAN GILS, L.G.M.; SCHOLZ, K.E.; WEST. C.E. (1983b): Serum cholesterol levels of calves and rabbits fed milk replacers containing skim milk powder or soybean protein concentrate. Nutr. Rep. Int. **27**, 757-764.
BEYNEN, A.C.; WEST, C.E. (1981): The distribution of cholesterol between lipoprotein fractions of serum from rabbits fed semipurified diets containing casein and either coconut oil or corn oil. Z. Tierphysiol. Tierernährg. u. Futtermittelkde. **46**, 233-239.
CARROLL, K.K. (1971): Plasma cholesterol levels and liver cholesterol biosynthesis in rabbits fed commercial or semisynthetic diets with or without added fats or oils. Atherosclerosis **13**, 67-76.
CLARKSON, T.B.; LOFLAND, H.B.; BULLOCK, B.C.; GOODMAN, H.O. (1971): Genetic control of plasma cholesterol. Studies on squirrel monkeys. Arch. Pathol. **92**, 37-45.
JACOBSON, N.L.; RICHARD, M.; BERGER, P.J.; KLUGE, J.P. (1974): Comparative effects of tallow, lard and soybean oil, with and without supplemental cholesterol, on growth, tissue cholesterol and other responses of calves. J. Nutr. **104**, 573-581.
KATAN, M.B.; BEYNEN, A.C.; DE VRIES, J.H.M.; NOBELS, A. (1986): Existence of consistent hyper- and hyporesponders to dietary cholesterol in man. Amer. J. Epidemiol. **123**, 221-234.
KEYS, A.; ANDERSON, J.T.; GRANDE, F (1965): Serum cholesterol response to changes in the diet. I. Iodine value of dietary fat versus 2 S-P. Metabolism **14**, 747-758.
Lipid Research Clinics Program (1984): The lipid clinics coronary primary prevention trial results. Parts 1 and 2. J. Amer. Med. Assoc. **251**, 351-374.
ROBERTS, D.C.K.; WEST, C.E.; REDGRAVE, T.G.; SMITH, J.B. (1974): Plasma cholesterol concentration in normal and cholesterol-fed rabbits. Atherosclerosis **19**, 369-380.
TERPSTRA, A.H.M.; SANCHEZ-MUNIZ, F.J.; WEST, C.E.; WOODWARD, C.J.H. (1982): The density profile and cholesterol concentration of serum lipoproteins in domestic and laboratory animals. Comp. Biochem. Physiol. **71B**, 669-673.

VAN DER MEER, R.; SCHÖNINGH, R.; DE VRIES, H. (1985): The phosphorylation state of casein and its differential hypercholesterolemic effect in rabbits and rats. In: Cholesterol Metabolism in Health and Disease: Studies in the Netherlands (Editors: BEYNEN, A.C.; GEELEN, M.J.H.; KATAN, M.B.; SCHOUTEN, J.A.). Ponsen & Looijen, Wageningen. Pp. 151-157.

VAN RAAIJ, J.M.A.; KATAN, M.B.; HAUTVAST, J.G.A.J.; HERMUS, R.J.J. (1981): Effects of casein versus soy protein diets on serum cholesterol and lipoproteins in young healthy volunteers. Amer. J. Clin. Nutr. 34, 1261-1271.

VAN ZUTPHEN, L.F.M.; DEN BIEMAN, M.G.C.W. (1981): Cholesterol response in inbred strains of rats, Rattus norvegicus. J. Nutr. 111, 1833-1838.

VAN ZUTPHEN, L.F.M.; DEN BIEMAN, M.G.C.W. (1983): Genetic control of plasma cholesterol response in the rat. J. Hered. 74, 211-212.

VAN ZUTPHEN, L.F.M.; FOX, R.R. (1977): Strain differences in response to dietary cholesterol by JAX rabbits: Correlation with esterase patterns. Atherosclerosis 28, 435-446.

WAGNER, W.B.; CLARKSON, T.B. (1974): Mechanisms of the genetic control of plasma cholesterol in selected lines of show racer pigeons. Proc. Soc. Exp. Biol. Med. 145, 1050-1057.

WEST, C.E.; BEYNEN, A.C. (1986): Are there atherogenic dietary proteins? Verhandl. Deutsch. Gesellsch. Innere Med. 92, 666-673.

Author's addresses:
A.C. Beynen, Department of Laboratory Animal Science, State University, P.O. Box 80.166, 3508 TD Utrecht, and Department of Human Nutrition, Agricultural University, P.O. Box 8129, 6700 EV Wageningen, The Netherlands.

The domestic chicken *(Gallus domesticus)* as an organism for the study of chromosomal aberrations[a]

N. S. FECHHEIMER

Introduction

The reproductive processes of man and domestic animals are subject to errors resulting in production of gametes and conceptuses with abnormal chromosomal complements. Observations of the chromosomes of human spontaneous abortions during the first trimester of pregnancy revealed 60% to be afflicted with an abnormal chromosome complement (BOUÉ et al. 1975). More recent studies of human chromosomes during maturation divisions of gonocytes or mitosis of the first or second cleavage division indicated that many additional chromosomal aberrations occur that are lethal at such early stages that the pregnancy is not recognized prior to embryonic mortality and abortion (MARTIN et al. 1986). Other classes of chromosomal abnormalities permit pregnancy to proceed to term but cause various stigmata of the stillborn or live-born neonates (HOOK 1982). The occurrence of chromosomal abnormalities is the most important single cause of pregnancy failure in man affecting at least 10% of all fertilized eggs, and contributes significantly to the burden of high rates of embryonic mortality of laboratory and farm animals (HARE & SINGH 1979).

If the pervasive occurrence of pregnancy wastage and its associated human suffering and economic costs is to be alleviated, it is necessary to delineate the mechanisms by which the errors occur and the causal factors of importance in producing chromosomally aberrant gametes and zygotes. These questions have been approached in two ways. Inferences can be drawn from data accumulated in numerous studies of human material, but the procedure is slow, costly and the tentative conclusions relate only to causal factors for which ancillary information is collected for retrospective analyses (for review see BOUÉ et al. 1985). Secondly, various aspects of the problem have been extensively studied in small, laboratory mammals, notably the mouse, rat, hamster, and rabbit (reviewed by DYBAN & BARANOV 1987). A third source of information bearing directly on the question of the etiology of chromosomal abnormalities in embryos has been largely neglected or thought to be not relevant to the situation in man or mammalian livestock. Embryos of the domestic chicken represent a particularly useful model system and the accumulated data are more extensive than that for any other organism except the laboratory mouse.

It is the aim of this paper to relate the approaches that have been taken using the early chick embryo, and to review briefly what has been learned that is applicable to the unanswered questions of etiology of chromosomal abnormalities in mammals.

The chicken karyotype

The normal chicken karyotype presents 39 pairs of chromosomes (2N = 78) of which nine pairs, including the gonosomes, are of sufficient size and distinctive morphology as to be individually distinguishable without the necessity of using banding procedures. The remaining 30 pairs are small, indistinguishable from one another at mitotic and meiotic metaphases, contain relatively large blocks of heterochromatin and are not readily amenable to cytogenetic analysis. The sex chromosomes comprise the Z, a metacentric element of about the same length as autosome number 4, and the W, a submetacentric element equivalent in length to autosome number 7. The W is largely heterochromatic, presenting a very prominent C-band, and fluoresces brightly when stained with quinacrine mustard, or with acradine orange after incorporation of BrdU (5-bromodeoxyuridine; R-banding). The female is heterogametic (ZW) and the male homogametic (ZZ), a situation which is advantageous when analyses are made for the source of extraneous or absent genomes in euploid (haploid and polyploid)

---

[a] *A glossary of cytogenetic terms is appended.*

embryos, and for the origins of sex chromosome aneuploidy. Descriptions of the karyotypes following application of various banding procedures (G-, Q-, C-, R-, NOR) have been published (WANG & SHOFFNER 1974, CARLENIUS et al. 1981). Each procedure reveals distinctive features of the larger chromosomes (macrochromosomes) but none have been very useful in identifying the small, microchromosomes. Silver staining, however, reveals one of the microchromosomes (number 15 to 18) as the carrier of the nucleolus organizing region (BLOOM & BACON 1985) and staining with distamycin A and chromomycin A has been used to identify several of the larger microchromosomes (AUER et al. 1987).

Meiotic chromosomes from testis of mature males are relatively easily analyzed at diakinesis - metaphase of the first meiotic division and metaphase of secondary spermatocytesispermatocytes, i.e. second meiotic metaphase (POLLOCK & FECHHEIMER 1978).

The large number of microchromosomes that cannot be individually identified, or even accurately counted in routine preparations, are a distinct hindrance to some types of cytogenetic analysis. However, the system as a whole has many advantages, making it particularly favorable for fruitful study of the etiology of abnormalities of the karyotype of gametes and zygotes.

Attributes of the chicken for cytogenetic studies

In addition to its favourable karyotype which, with the exception of the large numbers of microchromosomes, has many features that are eminently suitable for most cytogenetic analyses, the chicken possesses other attributes that suit it admirably for such work.

1) It has been widely used for study of embryological development (ROMANOFF 1960, 1972), physiology of reproduction, and genetics (HUTT 1949). As a consequence its normal biological properties are well known.

2) The gemeration interval of about one year, although not so short as that of most of the laboratory mammals, is still short enough to enable construction of genetically useful stocks and to conduct multiple-generation genetic analyses, such as selection experiments. Another gallinaceous bird, the Japanese quail (*Coturnix coturnix japonica*), has most of the other properties of the chicken and has the added advantage of a generation interval of about four months, comparable to that of the laboratory mouse.

3) A hen can be expected to produce 20 to 25 eggs per month in the first year of sexual maturity and if necessary can be recycled to produce eggs for a second year. Therefore, it is possible to collect large numbers of full-sib or half-sib embryos for cytogenetic analysis and still have the unimpaired dam available to produce eggs to reproduce.

4) Chromosome analysis successfully completed of up to 90% of embryos at one day of incubation (primitive streak stage), a stage prior to which there is very little if any mortality of genetically unbalanced zygotes. Furthermore, all ova that are ovulated are accounted for as eggs produced, and the eggs can be stored for 7 to 10 days before processing them for karyological observation without loss of developmental capacity.

5) The cost of producing eggs is low. Under semi-commercial conditions they can be had for about $10 (U.S.) per 100 eggs.

6) A number of stocks and strains are readily available including inbred lines, genetic and chromosomal markers such as translocations and inversions, lines selected divergently over many generations for a number of different characters, and random-bred stocks having diverse genetic backgrounds and maintained as large populations (SOMES 1978).

7) Some chromosomal abnormalities, notably triploidy, trisomy and some types of euploid chimerism are not obligatorily lethal in embryonic or fetal stages. Therefore live chicks bearing these abnormalities can be hatched and studied.

Procedures used

Sources and preparation of cells for chromosomal analysis

The methods applied to arrest mitosis at metaphase and treat cells to enhance the various morphological features of chromosomes are adaptations of those developed and widely used for observation of human and other mammalian chromosomes (SCHWARZACHER & WOLF 1974). Preparations can be made directly from biopsy specimens or pieces of tissues from sacrificed animals. Treatment of the animal with colchicine or other mitotic arresting agents increases the concentration of cells at metaphase. Alternatively tissue specimens such as bone marrow flushed from long bones, liver, bursa of Fabricius of newly hatched chicks can be put into suspension in any standard tissue culture medium and incubated for one to two hours at 40°C before treatment with a standard hypotonic agent and fixative. Cells from the pulp of growing feathers have been extensively used in one laboratory as a source of mitotic metaphase cells (SHOFFNER et al. 1967). Pieces of testis, when handled in a manner similar to that described by EVANS and coworkers (1964) for the mouse, yield excellent preparations containing abundant numbers of spermatogonia at metaphase, primary spermatocytes at diakinesis-metaphase of the first meiotic imeiosisdivision and fewer numbers at metaphase of second meiotic division (POLLOCK & FECHHEIMER 1978).

The chromosomal complement of embryos at one day of incubation has been extensively studied (for review see FECHHEIMER 1981). The preparative procedure, given in detail by MILLER and coworkers (1971) involves treatment of the embryo *in ovo* with colchicine, removal of the blastoderm from the egg yolk, making a suspension of the embryonic cells in a balanced salt solution, and thereafter following procedures appropriate for cultures of lymphocytes. BLOOM (1974) has used extraembryonic tissue from eggs incubated for four days for studies of many chicken embryos.

Analysis of chromosomes from cells in culture, particularly lymphocytes which can be obtained from a small blood sample, is a procedure generally applied for study of human subjects. With minor modification the technique is directly applicable to birds. The essential difference is that for growth of bird cells in culture, incubation of the cultures must be at a temperature of 40° to 41°C (AU et al. 1975). Fibroblasts from small skin biopsies of chickens are also cultured successfully at 40° to 41°C and yield excellent preparations (ANSARI et al. 1986).

The presence of polyploid cells in samples of blood can be unambiguously detected and quantified by fluorescence flow cytometry (THORNE et al. 1987). Presumably this methodology could be applied to samples of embryonic cells enabling the screening of many embryos for the presence of euploid cells without the necessity for labor-intensive and tedious analysis with the microscope.

Chromosome banding procedures

Most of the techniques that were developed to reveal various aspects of the internal structure of chromosomes by displaying sequential series of bands on them at appropriate stages of contraction (GUSTAVSSON 1980) have been adapted for use with chickens as well as other birds. The methods are not so reliable, or the results so consistent, when applied to the chromosomes of chickens as they are hen used on mammalian chromosomes. Such inconsistency might well be related to he fact that bird chromosomes have only about one-half the DNA content of mammalian chromosomes and the deficiency is entirely of non-coding heterochromatin and repetitive sequences. It is thought that certain classes of abundant repetitive sequences are responsible for the distinctive banding patterns of choromosomes (KORENBERG & RYKOWSKI 1988). Nonetheless procedures for production of G, Q, R, C, and NOR bands can be applied to aid in the recognition of individual chromosomes or segments of them, although it is infrequently necessary to do so in experiments to study the etiology of heteroploidy in chicken embryos.

Production of marker chromosomes

To determine the origin of various types of euploidy occurring in embryos it is helpful to have either the maternal or paternal genome stamped with a homozygous marker chromosome. The most easily produced of marker chromosomes are reciprocal translocations or pericentric inversions. Both have been efficiently created by treatment of samples of spermatozoa with appropriate doses of X-rays (Wooster et al. 1977) or radiomimetic agents such as EMS (ethyl methane sulfonate) or TEM (triethylene melamine) (Shoffner 1972). Chicks are then hatched from eggs produced by hens inseminated with the treated semen and screened for the presence of readily detectable rearrangements. In one such experiment 19 translocations and one inversion were recovered from only 200 chicks derived from X-irradiated semen. All were fully fertile (Wooster et al. 1977). Some of the rearrangements were made homozygous by mating of heterozygous half-sibs produced from the original carrier of a rearrangement. A number of stocks heterozygous and homozygous for rearrangements are maintained at University of Minnesota and at the Ohio State University.

Types of chromosomal abnormalities observed in chick embryos

In a series of studies incorporating observation of more than 6,000 embryos summarized by Bloom (1974) and another series encompasing more than 8,000 embryos summarized by Fechheimer (1981) all the types of heteroploidy that occur in man and other mammals were observed. All levels of euploidy comprising haploidy and polyploidy from 3N to 8N, with the single exception of 7N, were seen in various numbers of embryos. In addition trisomy and monosomy for each of the eight macrochromosomes occurred in frequencies high enough to enable inferences to be made regarding their source and causes. Double and multiple aneuploidy as well as aneuploid mosaicism were also observed. An unusual feature of embryos from some chicken strains was the relatively high incidence of euploid chimeras of types such as 1N/2N, 1N/3N, and 2N/2N and 2N/3N (Snyder et al. 1975; Fechheimer & Jaap 1978, 1980). In addition some lines were characterized by the occurrence of 2N/4N mosaics.

Because such a large sample of chick embryos has been observed and the sample includes representatives from a number of different strains in which the various types of heteroploidy occurred at very divergent frequencies, it has been possible to infer where in the reproductive cycle the errors occur resulting in each type of abnormality. The inferences thus made are in accord, for the most part, with what is known about the origins of similar abnormalities in one or more mammalian species.

Origins of abnormalities in chick embryos

The use of marker chromosomes to indicate the parental source of entire haploid genomes increases vastly the ability to detect the origin of particular types of abnormalities. The critical experiments were with embryos whose dams had normal karyotypes and whose sires were homozygous for an easily recognized marker chromosomes. In one experiment semen from several cockerels, each of which was homozygous for a different marker, was mixed and used to inseminate hens. The experiment enabled the detection of otherwise undetectable types of abnormalities such as 1N/1N and 2N/2N. In the following paragraphs what is known about the origins of each type of abnormality is summarized. More complete arguments and more thorough statements of the evidence in support of the conclusions is given in the papers cited, especially that of Fechheimer (1981).

Origin of triploidy (3N)

The gonosomal complement of 3N birds can be ZZZ, ZZW, or ZWW. Because the normal diploid female is ZW, the type ZWW must be the result of a digynic egg fertilized by a single spermatozoon. Furthermore ZZ and WW digynic eggs should be produced in equal numbers when the second meiotic division is suppressed. In an analysis of 147 triploid embryos there were 65 ZZZ, 26 ZZW, and 56 ZWW indicating that about 75% resulted from dygynic oocytes resulting from failure of second meiotic division. The remainder originated from digyny resulting from failure of first meiotic division and from dispermy (FECHHEIMER 1981). This conclusion was confirmed by the experiments with marker chromosomes in which all 20 triploids contained only a single set of paternal chromosomes.

Origin of pentaploidy (5N)

The two pentaploids recovered in the experiments involving marker chromosomes contained only a single paternal genome. Apparently they were derived from an oocyte rendered tetraploid by the failure of both meiotic divisions. The 4N oocytes were fertilized by single sperm.

Origin of tetraploidy (4N) and diploid/tetraploid (2N/4N) mosaicism

The 4N chromosome complement of 2N/4N mosaic embryos in every case observed represented an exact doubling of the 2N cell line complement. This was true for sex chromosomes, marker chromosomes, aneuploidy, deletions or other structural aberration. The 4N cell line of such embryos therefore must arise from a failure of cytokinesis at mitosis of an early cleavage division, following normal fertilization. Pure tetraploid embryos presumably are formed from failure of the first cleavage division. The same error, i.e. failure of cytokinesis at mitosis, is thought to be the source of the infrequently occurring 3N/6N mosaic embryos and of a small proportion of 1N/2N embryos.

Origin of haploid/diploid (1N/2N), haploid/triploid (1N/3N), pure haploid (1N) and other euploid chimeric embryos with 1N cell line

From the experiments in which marker chromosomes were used 77 embryos were recovered that were pure haploid or in which, more commonly, a haploid cell component was present as well as one or more other euploid cell lines. The latter embryos were largely 1N/2N but other types such as 1N/2N/4N and 1N/3N were also recovered. In every case, the haploid cell line was androgenetic, i.e. originated from a spermatozoon. The predominant type, 1N/2N, therefore arises as a result of dispermy. One sperm enters the egg and fuses with the egg pronucleus in a normal way to give rise to the 2N component cell line. A second spermatozoon enters the egg, does not engage in syngamy but simply proliferates by mitosis to yield the haploid cell component of the embryo. Pure haploid embryos result when pronuclear fusion does not occur; the female pronucleus presumably disintegrates or in any case does not divide, but the sperm pronucleus undergoes mitosis to give rise to an androgenetic haploid embryo. Occasionally, in about 10% of cases, more than one supernumerary sperm form independent 1N cell lines resulting in embryos of the form 1N/1N and 1N/1N/2N. This is known from the experiment where semen from cockerels homozygous for different marker chromosomes was mixed before insemination of hens (FECHHEIMER & JAAP 1980).

Origin of diploid/triploid (2N/3N) chimeric embryos

This type of heteroploidy, although it occurs infrequently in chick embryos (about 1.5 per 1,000 embryos) is of great interest because it is not entirely lethal and has been observed in all laboratory mammals as well as man, cattle, cat, horse and mink (see review in FECHHEIMER et al. 1983). A total of 15 chicks has been reported. A number of different mechanisms have been proposed as origins. Six of these mechanisms seem to be biologically feasible and each has an expected array of sex chromosome complement combinations for the chimeric zygotes that it should yield. There are six such combinations; each of the diploid types ZZ and ZW can occur in combination with each of the three triploid types ZZZ, ZZW, and ZWW. The sex chromosome complements of 15 embryos was compared to the expected array from each of the proposed six feasible mechanisms. Only one of the six could have yielded all of the embryos observed, and therefore was presumed to be the mechanism responsible for a great majority, if not all, of the aberrant embryos. The favored mechanism involves the fertilization by one or two spermatozoa of each pronucleus resulting from maturation of binucleated oocytes. Each of the two nuclei of the binucleated oocyte can have the first or second meiotic division fail, rendering its pronucleus as diploid. The same analysis was applied to a sample of 15 postimplantation mink embryos with the same result. All could have arisen *via* the same mechanism (FECHHEIMER et al. 1983), indicating the value of the comparative approach.

Origins of diploid/diploid (2N/2N) chimerism

This abnormality occurs with about the same frequency as 2N/3N chimerism when it is ascertained in experiments using marker chromosomes. When marker chromosomes are not used only half of such embryos can be detected. It is not possible to apply an analysis similar to that used for 2N/3N to infer its mechanism of origin. It occurs however in the same strains of chicken and mink in which 2N/3N occurs, and could reasonably be thought also to originate from binucleated oocytes.

Origin of aneuploidy

Of the many types of sex chromosome aneuploidy that might be expected only two have been recovered, ZZW and ZO. They occurred in roughly equal numbers of 9 and 6 respectively. The mechanism most likely to produce such an outcome is nondisjunction of the sex chromosome bivalent at first meiotic division of the oocyte, yielding pronuclei bearing ZW or O sex chromosomes in equal numbers. Fertilization by normal sperm would produce ZZW and ZO zygotes, as observed.

Autosomal aneuploidy is encountered relatively infrequently in chick embryos because it can be cytologically recognized for only one-fourth of the 38 pairs of autosomes (8 macrochromosome pairs of the total autosomal complement of 38 pairs).

Owing to the paucity of data it is not possible to estimate the contribution of nondisjunction during meiosis of oocytes and spermatocytes to the overall number of aneuploid embryos. It is known however, from analysis of half-sib families that both occur.

Etiology for aberrations

It is not yet possible to identify specific causal factors for any particular type of chromosomal abnormality, probably because the system is not so simple that single factors are agents for specific errors of the reproductive process resulting in production of a single type of chromosomally abnormal embryo. Nonetheless, retrospective analysis of accumulated data, as well as results from experiments made to test hypothesis, give strong indication of cause and effect relationships between various suspected causes and the occurrence of particular types of abnormalities. Of special interest is the evidence suggesting an important genetic influence on the frequency of occurrence of the more predominant forms of heteroploidy.

Etiology of triploidy (3N)

Spontaneously occurring triploidy in the chicken is usually the result of suppression of second meiotic division of the oocyte. The error occurs more frequently when ovulation is mistimed, yielding eggs that are ovulated prematurely, or are senescent owing to delayed ovulation. Thus triploidy is recovered more frequently from eggs produced early in the laying cycle than from those produced later in the cycle when a regular daily cycle of ovulation is established (MONG et al. 1974). Furthermore, the incidence of triploidy in embryos recovered from double-yolked eggs was very high (7%) compared to those from normal single yolked-yolked eggs (0%) produced by the same hens (LEE et al., unpublished).

There appears to be a strong genetic influence affecting the incidence of triploidy. A line of meat-type chickens produced four times more 3N embryos than an egg-producing strain maintained in a similar environment (FECHHEIMER 1981). Within lines of chickens and *Coturnix* certain hens yield significantly more triploid embryos than others (BLOOM 1974). In a sample of 600 *Coturnix* embryos from 36 hens, 15 were triploid and all were the progeny of three hens (DE LA SENA et al., unpublished). It is not yet known if the genetic effect affects the occurrence of triploidy by interfering with the regularity of ovulation, or causes failure of the second meiotic division by other means. Thorough analysis of a line successfully selected for a increased incidence of triploid should yield useful answers to these questions (THORNE et al. 1987).

Etiology of haploidy (1N) and haploid/diploid (1N/2N) chimerism

Large differences among different genetic stocks in the frequency of occurrence of these chimeric embryos has been noted (BLOOM 1974, FECHHEIMER 1981). In one such comparison a broiler line produced 20 times more haploid chimeric embryos than an egg laying line housed and cared for in a similar environment. Crosses between the lines yielded an intermediate frequency that depended upon the genotype of the female parent of the embryos, the male parent having no effect (SNYDER et al. 1975). This result was surprising because the haploid component cell line of 1N/2N chimeric embryos are known to be derived from supernumerary spermatozoa that proliferate by mitosis within the egg. Haploid chimeric embryos are not distributed randomly among the embryos produced by hens of any particular line of chickens. They exhibit a clustering, having high frequency in eggs of some dams, whereas eggs from other hens never contain embryos with this aberration. In both chicken and *Coturnix*, the frequency of haploid chimeras is three to four times higher in lines selected for rapid growth than in random-bred controls, or lines selected for slow growth (REDDY & SIEGEL 1977, WOLOWODIUK et al. 1985). Whether this association represents a genetic correlation or is simply a spurious outcome of genetic drift, is not clear. The evidence as a whole indicates, however, that the primary causative factor of haploid chimerism is the dam's genotype. It is of interest to determine if the genetic influence is exerted simply by a faulty block to polyspermy, or if the eggs of susceptible dams contain a factor that elicits, or at least does not suppress, mitosis of spermatozoa in the blastoderm.

Etiology of 2N/4N mosaicism

This abnormality occurs sporadically, at differing frequencies in all lines of birds investigated. In most lines its incidence is greater than any other form of heteroploidy except haploid chimerism. It exhibits an extreme tendency to occur in certain families at high frequency. In one sibship of meat-type chickens 15 of 23 embryos were 2N/4N. A full-sib male of this family sired five 2N/4N among a progeny of 11 embryos (MILLER et al. 1971, 1976). From this and extensive further evidence it is apparent that a gene or genes segregating in some populations of birds, particularly those selected for rapid growth, suppresses cytokinesis of an early cleavage division resulting in a doubling of the chromosome complement in the cell line derived from the affected cell or cells.

Etiology of aneuploidy

Pure autosomal aneuploidy, trisomy and monosomy, is recovered relatively infrequently in chick embryos, in part no doubt because only one fourth of them can be detected. For this reason it has been impossible to accumulate samples of sufficient size for etiologic analysis. In man the incidence of trisomy increases among conceptions of older mothers. Chickens are not kept past two to three years of age although their natural life-span is in excess of 10 years. Therefore one cannot use the chicken to seek the reason for the large increase in meiotic nondisjunction of females of advanced age. What has been noted in chicken embryos is a remarkable clustering within families of full-sib or half-sib embryos. In many of such families moreover, it has been found that the several affected embryos are trisomic or monosomic for the same chromosome. Thus familial clusters have been observed for aneuploidy of chromosomes 1, 2 and 4 (FECHHEIMER 1981). It would appear that chromosomes with mutant centromeres are segregating at low frequencies in populations of domestic chickens. The mutant centromeres exhibit the tendency for nondisjunction at first or second meiosis in both males and females.

Monosomy results from chromosome lagging as well as from nondisjunction. When either event occurs at mitosis of a cleavage division, its occurrence will be indicated by the presence of aberrant cell lines in embryos, both monosomic and trisomic cells in the case of nondisjunction, but only monosomic cells in the case of lagging. A high incidence of diploid/hypodiploid mosaic embryos was observed in a random-bred population of domestic turkeys. Only three of the five recognizable autosomes were monosomic in any of the aberrant cell lines. In a parallel line selected from the random bred line for rapid growth, the incidence of similarly aberrant embryos was significantly reduced to only half the incidence in the control line (FECHHEIMER & NESTOR 1988). One reasonable interpretation is that chromosomes with a tendency for lagging at mitotic metaphase, that were present in the control population, were largely eliminated during the process of selection of the divergent line. If this is correct, it is a clear indication of mutant centromeres affecting chromosome movement at mitosis but not at meiosis.

Disjunctional behavior of multivalents

Another aspect of human and animal cytogenetics which can be profitably studied in the chicken is that involving the pairing behavior and disjunctional properties of chromosomes bearing structural aberrations, particularly translocations and inversions. In individuals heterozygous for a translocation, multivalents involving the translocated chromosomes and the normal homologues are formed at prophase of the first meiotic division. Segregation can then take a variety of forms only one of which yields genetically balanced gametes. All others yield genetically unbalanced gametes that contain duplicated or deleted whole chromosomes or chromosome segments. Such gametes might, or might not, be equally capable of engaging in fertilization but bring to the zygote a chromosome complement that cannot sustain normal development (FORD & CLEGG 1969).

The problem is not inconsequential because about 2 to 5 per 1,000 newborn babies have a balanced rearrangement (HOOK & HAMERTON 1977) and certain breeds of livestock have relatively high frequencies of a particular type of translocation, centric fusions (ELDRIDGE 1985). It is of considerable importance both in medical genetics and animal breeding to be able to estimate the effect on reproductive fitness of heterozygosity for structural rearrangements. Although only a limited amount of work has been done with the chicken, it has become apparent that useful information, applicable to the situation in mammals, can be elicited from study of rearrangements in chickens.

Pairing properties of structurally altered chromosomes

The method for electron microscopy of synaptonemal complexes of spermatocytes and oocytes was adapted for use with chickens (SOLARI 1977, KAELBLING & FECHHEIMER 1983a). The pairing behavior of several reciprocal translocations and one inversion have been extensively observed (KAELBLING & FECHHEIMER 1983b, 1985, SOLARI et al. 1989). This work has established that chicken chromosomes involved in rearrangements behave in a manner similar to those of man (CHANDLEY 1988) and small mammals and therefore provide a valid, low cost model system for study of various phenomena associated with pairing and subsequent adjustment of chromosome segments in first meiotic prophase.

Segregational properties of rearrangements

Analysis of transmission ratios of the various gametic products of segregation of quadrivalents provided interesting and unexpected results. The experiments involved karyotyping of embryos resulting from matings between sires or dams that were heterozygous for a translocation and karyologically normal mates. In every instance in which the female parent was heterozygous the transmission ratios were significantly different from the theoretical expectation. The difference could be explained in each instance by a lagging at first meiotic metaphase of one of the two translocation chromosomes (DINKEL et al. 1979). On the other hand, embryos derived from heterozygous sires, except in one instance, exhibited no deviations from the expected 1:1 ratio of complementary gametic products. Segregation or transmission of gametic products of a quadrivalent were seen to be different in males and females.

In the single instance in which transmission ratios of products of segregation from males was deviant, analysis of karyotypes of secondary spermatocytes indicated that segregational anomalies were occurring at first meiotic anaphase. Thus the complementary products of segregation were not produced in the expected 1:1 ratio. Further, the sperm that fertilized eggs represented a significantly different array than that seen in secondary spermatocytes. There was clear indication of differential survival or functional capacity of sperm bearing different chromosomal complements (BONAMINIO & FECHHEIMER 1989).

Summary

Early embryonic mortality occurs with high frequency in man and domestic animals. Although many factors have been implicated as etiological agents, a high proportion (about 50%) of human spontaneous abortuses bear an abnormal chromosomal complement. The abnormal complements result from errors at all stages of the reproductive process including germ cell proliferation and meiotic divisions of both spermatocytes and oocytes, fertilization, and mitosis of early cleavage divisions. Extensive studies with human conceptuses and embryos of laboratory mammals have been made to detect the origins of the karyological abnormalities and delineate the causal factors for occurrence of the errors.

The domestic chicken possesses many attributes that suit it admirably for similar work, and the cost of experiments with chick embryos is relatively low. The availability of strains with high frequency of particular types of abnormalities, and of others bearing readily detectable marker chromosomes enable the conduct of definitive experiments to identify the source of each of the types of heteroploidy that occurs. From observation of large samples of half-sib and full-sib embryos, as well as of embryos from a number of different strains and their crosses, it has become apparent that genetic factors are of importance in causing the errors resulting in the occurrence of heteroploid embryos. Other non-genetic factors have been identified as causal agents of particular types of abnormalities.

Embryos with abnormal chromosome complements also occur when one parent is heterozygous for a chromosomal rearrangement. In the case of translocations, multivalents are formed at first meiotic division. Segregation from multivalents results in production of both normal gametes and gametes bearing both duplication and deletion of segments of chromosomes, or of whole chromosomes. Lines of birds bearing translocations are being used to elucidate the factors that influence the segregation patterns of translocation multivalents. Males heterozygous for translocations are also used as a source of both normal and genetically unbalanced spermatids and spermatozoa to study their relative functional competencies.

Acknowledgement

Gratitude is extended to Professor G. F. Stranzinger and the staff of the Institute of Animal Science, Federal Institute of Technology, Zürich, Switzerland for the hospitality tendered during the course of a leave, when this paper was written.

References

ANSARI, H. A.; TAKAGI, N.; SASAKI, M. (1986): Interordinal conservatism of chromosome banding patterns in *Gallus domesticus (Galliformes)* and *Melopsittacus undulatus (Psittaciformes)*. Cytogenet. Cell Genet. 43, 6-9.
AU, W.; FECHHEIMER, N. S.; SOUKUP, S. (1975): Identification of the sex chromosomes in the bald eagle. Can. J. Genet. Cytol. 17, 187-191.
AUER, H.; MAYR, B.; LAMBROU, M.; SCHLEGER, W. (1987): An extended chicken karyotype including the NOR chromosome. Cytogenet. Cell Genet. 45, 218-221.
BLOOM, S. E. (1974): The origins and phenotypic effects of chromosome abnormalities in avian embryos. Proc. 15th World's Poultry Congress. McGregor and Warner, Washington, D.C. Pp. 316-320.
BLOOM, S. E.; BACON, L. D. (1985): Linkage of the major histocompatibility (B) complex and the nucleolar organizer in the chicken. J. Hered. 76, 146-154.
BONAMINIO, G. A.; FECHHEIMER, N. S. (1988): Segregation and transmission of chromosomes from a reciprocal translocation in *Gallus domesticus* cockerels. Cytogenet. Cell Genet. 48, 193-197.
BOUÉ, A.; BOUÉ, J.; GROPP, A. (1985): Cytogenetics of pregnancy wastage. Adv. Human Genet. 14, 1-57.

Boué, J.; Boué, A.; Lazar, P. (1975): Retrospective and prospective epidemiological studies of 1500 karyotyped spontaneous human abortions. Teratology **12**, 11-26.

Carlenius, C.; Ryttman, H.; Tegelstroem, H.; Jansson, H. (1981): R-, G-, and C-, banded chromosomes in the domestic fowl *(Gallus domesticus)*. Hereditas **94**, 61-66.

Chandley, A. C. (1988): Meiosis in man. Trends Genet. **4**, 79-84.

Deboer, L. E. M.; Degroen, T. A. G.; Frankenhuis, M. T.; Zonnenveld, A. J.; Sallevelt, J.; Belterman, R. H. R. (1984): Triploidy in *Gallus domesticus* embryos, hatchlings, and adult intersex chickens. Genetica **65**, 83-87.

Dinkel, B. J.; O'Laughlin-Phillips, E. A.; Fechheimer, N. S.; Jaap, R. G. (1979): Gametic products transmitted by chickens heterozygous for chromosomal rearrangements. Cytogenet. Cell Genet. **23**, 124-136.

Dyban, A. P.; Baranov, V. S. (1987): Cytogenetics of Mammalian Embryonic Development. Clarendon Press, Oxford.

Eldridge, F. E. (1985): Cytogenetics of Livestock. AVI Publ. Co., Westport, CT. Pp. 122-143.

Evans, E.P.; Breckon, G.; Ford, C.E. (1964): An air drying method for meiotic preparation from mammalian testes. Cytogenetics **3**, 295-298.

Fechheimer, N. S. (1981): Origins of heteroploidy in chicken embryos. Poult. Sci. **60**, 365-371.

Fechheimer, N. S.; Isakova, G. K.; Belyaev, D. K. (1983): Mechanisms involved in the spontaneous occurrence of diploid-triploid chimerism in the mink *(Mustela vison)* and chicken *(Gallus domesticus)*. Cytogenet. Cell Genet. **35**. 238-243.

Fechheimer, N. S.; Jaap, R. G. (1978): The parental sources of heteroploidy in chick embryos determined with chromosomally marked gametes. J. Reprod. Fert. **52**, 141-146.

Fechheimer, N. S.; Jaap, R. G. (1980): Origins of euploid chimerism in embryos of *Gallus domesticus*. Genetica **52/53**, 69-72.

Fechheimer, N. S.; Nestor, K. E. (1988): Chromosomal abnormalities in turkey embryos from a control line and a line selected for rapid growth. In: Proc. 8th European Colloquium on Cytogenetics of Domestic Animals. Bristol, England, 19-22 July, 1988. (Editor: Long, S.E.), (in press).

Ford, C. E.; Clegg, H. M. (1969): Reciprocal translocations. Brit. Med. Bull. **25**, 10-114.

Gustavsson, I. (1980): Banding techniques in chromosome analysis of domestic animals. Adv. Vet. Sci. Comp. Med. **24**, 245-289.

Hare, W. C. D.; Singh, E. L. (1979): Cytogenetics in Animal Reproduction. Commonwealth Agricultural Bureaux, Slough, England.

Hook, E. B. (1982): Contribution of chromosomal abnormalities to human morbidity and mortality. Cytogenet. Cell Genet. **33**, 101-106.

Hook, E. B.; Hamerton, J. L. (1977): The frequency of chromosome abnormalities detected in consecutive newborn studies - Differences between studies - Results by sex and by severity of phenotypic involvement. In: Population Cytogenetics (Editors: Hook E.B.; Porter I.A.). Academic Press, New York. Pp. 63-79.

Hutt, F. B. (1949): Genetics of the Fowl. McGraw-Hill, New York, NY.

Kaelbling, M.; Fechheimer, N. S. (1983a): Synaptonemal complexes and the chromosome complement of domestic fowl, *Gallus domesticus*. Cytogenet. Cell Genet. **35**, 87-92.

Kaelbling, M.; Fechheimer, N. S. (1983b): Synaptonemal complex analysis of chromosome rearrangement in domestic fowl, *Gallus domesticus*. Cytogenet. Cell Genet. **36**, 567-572.

Kaelbling, M.; Fechheimer, N. S. (1985): Synaptonemal complex analysis of a pericentric inversion in chromosome 2 of domestic fowl, *Gallus domesticus*. Cytogenet. Cell Genet. **39**, 82-86.

Korenberg, J. R.; Rykowski, M. C. (1988): Human genome organization: Alu, Lines, and the molecular structure of metaphase chromosome bands. Cell **53**, 391-400.

Lee, K. H. (1985): Heteroploidy in early chick embryos from a line selected for an increase in capacity to produce double-yolked eggs. M.Sc. Thesis, The Ohio State University, Columbus, OH.

Martin, R. H.; Mahhadevan, M. M.; Taylor, P. J.; Hildebrand, K.; Long-Simpson, L.; Peterson, D.; Yamomoto, J.; Fleetham, J. (1986): Chromosomal analysis of unfertilized human oocytes. J. Reprod. Fert. **78**, 673-678.

Miller, R. C.; Fechheimer, N. S.; Jaap, R. C. (1971): Chromosome abnormalities in 16- to 18-hour chick embryos. Cytogenetics **10**, 121-136.

Miller, R. C.; Fechheimer, N. S.; Jaap, R. G. (1976): Distribution of karyotype abnormalities in chick embryo sibships. Biol. Reprod. **14**, 549-560.

Mong, S. F.; Snyder, M. D.; Fechheimer, N. S.; Jaap, R. G. (1974): The origin of triploidy in chick *(Gallus domesticus)* embryos. Can. J. Genet. Cytol. **16**, 317-322.

Pollock, B. J.; Fechheimer, N. S. (1981): Variable C-banding pattern and a proposed C-band karyotype in *Gallus domesticus*. Genetica **54**, 273-279.

Pollock, D. L.; Fechheimer, N. S. (1978): The chromosomes of cockerels *(Gallus domesticus)* during meiosis. Cytogenet. Cell Genet. **21**, 267-281.

REDDY, P. R. K.; SIEGEL, P. B. (1977): Chromosomal abnormalities in chickens selected for high and low body weight. J. Hered. **68**, 253-256.
ROMANOFF, A. L. (1960): The Avian Embryo: Structural and Functional Development. Macmillan, New York, NY.
ROMANOFF, A. L. (1972): Pathogenesis of the Avian Embryo. Wiley Interscience, New York, NY.
SCHWARZACHER, H. G.; WOLF, U. (Editors) (1974): Methods in Human Cytogenetics. Springer-Verlag, New York, Heidelberg, Berlin. Pp. 295.
SHOFFNER, R. N. (1972): Mutageneic effects of triethylene melamine (TEM) and ethyl methanesulfonate (EMS) in the chicken. Poult. Sci. **51**, 1865.
SCHOFFNER, R. N.; KRISHAN, A.; HAIDEN, G. J.; BAMMI, R.; OTIS, J. (1967): Avian chromosome methodology. Poult. Sci. **46**, 333-344.
SNYDER, M. D.; FECHHEIMER, N. S.; JAAP, R. G. (1975): Incidence and origin of heteroploidy, especially haploidy, in chick embryos from intraline and interline matings. Cytogenet. Cell Genet. **14**, 63-75.
SOLARI, A. J. (1977): Ultrastructure of the synaptic autosomes and the ZW bivalent in chicken oocytes. Chromosoma **64**, 155-165.
SOLARI, A. J.; FECHHEIMER, N. S.; BITGOOD, J. J. (1988): Pairing of ZW gonosomes and the localized recombination nodule in two Z-autosome translocations in *Gallus domesticus*. Cytogenet. Cell Genet. **48**, 130-136.
SOMES, R. G., Jr. (1978): Registry of Poultry Genetic Stocks. Storrs Agric. Exp. Sta. Bull No. 446. University of Connecticut, Storrs, CT, USA.
THORNE, M. L.; COLLINS, R. K.; SHELDON, B. L. (1987): Live haploid-diploid and other unusual mosaic chickens *(Gallus domesticus)*. Cytogenet. Cell Genet. **45**, 21-25.
WANG, N.; SHOFFNER, R. N. (1974): Trypsin G- and C- banding for interchange analysis and sex identification in the chicken. Chromosoma **47**, 61-69.
WOLOWODIUK, V. D.; FECHHEIMER, N. S.; NESTOR, K. E.; BACON, W. L. (1985): Chromosome abnormalities in embryos from lines of Japanese quail divergently selected for body weight. Genet. Sel. Evol. **17**, 183-190.
WOOSTER, W. E.; FECHHEIMER, N. S.; JAAP, R. G. (1977): Structural rearrangements of chromosomes in the domestic chicken: experimental production by X-irradiation of spermatozoa. Can. J. Genet. Cytol. **19**, 437-446.

Glossary of cytogenetic terms used

Androgenetic	Having origin from the male parent only, with no contribution from the female parent.
Aneuploidy	Chromosome number that is not an exact multiple of the characteristic number for the species.
Autosome	Any chromosome that is not a sex chromosome (gonosome).
C-band	Regions of chromosomes with special staining properties, that contain large blocks of heterochromatin.
Centric fusion	A type of translocation whereby two single armed chromosomes are joined at their centromeric ends to form a single biarmed chromosome.
Centromere	The localized region of chromosome to which spindle fibers are attached at mitosis and meiosis.
Chimera	An embryo or animal having cells with genetically different components. The component cell lines are derived from genetically different gametes.
Chromosome lagging	Delayed migration of a chromosome from the equatorial plate of metaphase to the poles, resulting in its exclusion from either of the daughter cells.
Dispermy	Penetration of an ovum by two spermatozoa.
Digyny	Presence of two maternal pronuclei in an ovum.
Euploidy	A chromosome number that is an exact multiple of the characteristic number for the species.

G-bands	Differentiation of segments of a chromosome into darkly and lightly staining regions by treatment with trypsin or other agents, followed by staining with Giemsa stain.
Gonosomes	Sex chromosomes (X and Y in most mammals, and Z and W in birds).
Haploid	Cell or organism containing the chromosome number characteristic of gametes (N).
Heteroploid	Cell or organism with a chromosome number different from that which is characteristic for the species.
Heterochromatin	Genetically inactive region of chromosome that has different staining properties than the remainder of the chromosome.
Hypodiploid	Chromosome complement differing from the normal diploid (2N) by the absence of one or more chromosomes.
Inversion	Rotation through 180° of an interstitial chromosome segment so that the sequence of genes it contains has a new position relative to the other genes of the chromosome.
Marker chromosome	One with markedly altered morphology by which it is readily identified and distinguished from others.
Monosomy	Chromosome complement differing from the number characteristic (2N) for the species by having one chromosome missing.
Mosaic	Embryo or organism composed of a mixture of genetically different cells that were derived from a single fertilized egg.
Multivalent	Configuration of more than two chromosomes in paired association at prophase and metaphase of first meiotic division.
N	Gametic chromosome number characteristic for the species. The haploid number of chromosomes.
Nondisjunction	Failure of chromatids or homologous chromosomes (first meiotic division) to separate at anaphase and move in opposite directions. The result is daughter cells that receive both and neither of the elements.
NOR	Nucleolar organizing region of particular chromosomes identified by a silver staining procedure.
Polyploid	Chromosome number that is an exact multiple, of three or greater, of the gametic number characteristic for the species.
Pentaploid (5N)	Chromosome number five times that of the gametic number characteristic for the species.
Q-bands	Differentiation of segments of a chromosome into brightly and dully fluorescing regions following staining with quinacrine stain.
R-bands	Differentiated segments of chromosomes that are darkly and lightly stained in reverse of the staining properties of G-bands, i.e. darkly stained G-bands are lightly stained R-bands and *vice versa*.
Synaptonemal complex	Proteinaceous structure that enables pairing of homologous chromosomes at prophase of first meiotic division.
Tetraploid (4N)	Chromosome number four times that of the gametic number characteristic for the species.
Translocation	Exchange of segments between chromosomes, yielding two monocentric chromosomes.
Triploid (3N)	Chromosome number three times that of the gametic number characteristic for the species.
Trisomy	Chromosome complement differing from the number characteristic for the species by having one additional chromosome.

Author's address:
N.S. Fechheimer, Department of Dairy Science, The Ohio State University, Columbus, OH 43210, U.S.A.

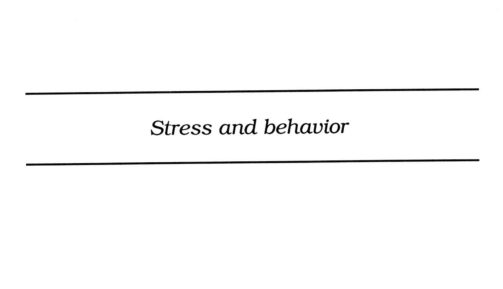

Stress: biological and biomedical considerations

V. PLIŠKA

The organism in its environment

The domain in which functions of a living system can be exercised is rather narrow. Physical and chemical processes in an organism require, for their completion, physicochemical conditions lying within sharply defined ranges. Keeping these conditions at their optimal values and, consequently, keeping the biological events in a steady state, is the task of a group of control processes vaguely termed "homeostatic mechanisms". To them one commonly attributes the regulation of blood flow, osmolarity, viscosity, ionic balances and pH of body fluids, regulation of temperature etc. The time course and the steady state of these processes are determined by the intrinsic state of the organism and by manifold environmental conditions.

Within a certain range, the effects of noxious stimuli (i.e. those leading to a violation of homeostasis and, frequently, to a disease condition) can be buffered by compensatory processes, so that a hypothetical threshold of "normality" is not overstepped. Even if this happens, any organism functioning regularly possesses mechanisms which cause the rearrangement of its elements in such a way that its basic functions may be maintained in new, albeit unfavorable conditions. This phase of existence is called "adaptation". Dependent upon the intensity of the noxious stimuli and their duration, the course of the homeostatic processes may turn back to the norm, into the homeostatic phase; in some other instances, the adaptation phase may become permanent. Only when both intensity and duration of a noxus become critical, an upper limit of the adaptive capacity is exceeded and detrimental changes within the organism become irreversible; functions of vital importance cannot be exercised, and collapse and death of the organism follow. This notion is schematically shown in Fig. 1.

Stress

According to the current nomenclature, the environmental stimuli which lead to an imbalance of homeostasis are called "stressors", and the corresponding defense reactions of the organism "stress". Hitherto the most important ideas on the phenomenon "stress" were published by H. SELYE in a "Nature" article of less than one printed page of length, in 1936. In his experiments on rats, several noxious agents like injury, cold, extensive physical exercise, toxic substances, etc., elicited similar pathologic symptoms which proceeded within three time phases. This notion is summarized in Fig. 2. In the course of the first two phases of persistent stressor action, the "alarm reaction" and the "resistance" ("adaptation") phase, nervous and hormonal pathways were involved. This regular pattern received the name "general adaptation syndrome", and is commonly abbreviated as "GAS" (SELYE & PENTZ 1943, SELYE 1950). Physiological processes in the third - "exhaustion" - phase are less well known. One can hypothesize that they occur in a state designated in the systems theory as "chaotic" (HAKEN 1978, HAKEN 1982).

Stressors described in the H. SELYE's communication were merely of experimental nature (they were extreme in their form and intensity) and by no means the only ones which lead to a similar pathological pattern. In fact, his experiments on rats in which the notion of GAS was formulated, were prompted by clinical experience with patients in early states of rather manifold acute, mostly infectious, diseases (SELYE 1967). Symptoms like headache, pain in the joints and extremities, disorders in the gastrointestinal tract, inappetentia, loss of body weight, increased urinary secretion of nitrogen bodies, phosphates, potassium, etc., are common in all patients and conditions. Due to their nonspecific nature, they are of little value for accurate diagnosis and only a minor significance has been ascribed to them. From the present viewpoint, they are signs of the alarm reaction that accompanies any pathogenic stressor. On the other hand, the endocrine, histological, biochemical and other changes, observed on rats in SELYE's experiments, scarcely occur with equal intensity in any

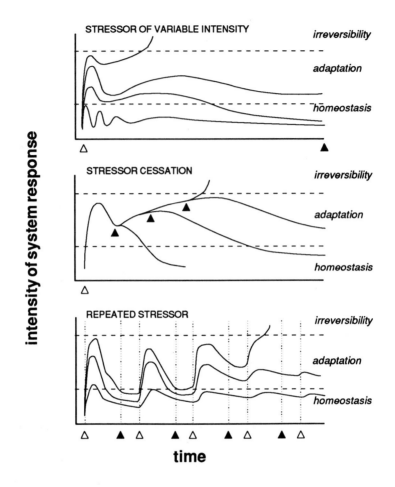

Fig. 1. Biological systems in response to stressors. Curves represent a hypothetic time course of a system function subjected to effect of a stressor (e.g., blood pressure, temperature, blood/tissue pH, concentrations of hormones, metabolites, etc.). It is assumed that, in the absence of stressors, this function is in the steady state. Minor perturbations caused by a weak and short-acting stressor can be buffered by current homeostatic processes (bottom zones of the diagrams). More vigorous and/or long lasting stressors activate additional compensatory mechanisms; the state of the system is transient but a return to an original, or a modified, steady state is possible; "damage" of the system caused by the stressor is reversible (middle zones of the diagrams). Effects of extremely strong and long lasting stressors are beyond a control of homeostatic and adaptation processes, changes of the system functions are irreversible (steady state cannot be renewed); the system undergoes a severe and irreparable damage. Upper panel visualizes effects of stressors of variable intensity upon a descriptor of the function in question (ordinate), middle panel effects of stressor duration, and lower panel effect of a repeated stressor. White and black triangles symbolize the beginning and end of stressor action, respectively.

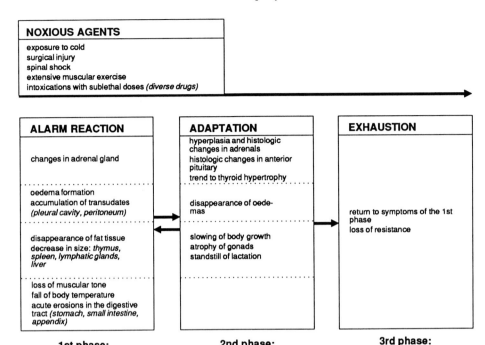

Fig. 2. Symptoms associated with individual phases of an experimentally elicited stress on rats (SELYE 1936). Stressors employed in the rat model are shown in the upper part of the diagram. Note that this profile is not necessarily uniform in all types of stress (INGLE 1952).

condition on human patients (INGLE 1952). Our own observations on stress-sensitive pigs do not document any major histological changes of the adrenal cortex (PLIŠKA et al., in preparation; cf. HARI, this monograph), thus indicating that also certain animal models deviate from the pattern of a stress response to experimental stressors of extreme intensity. Moreover, the pattern of GAS may be species dependent.

Stressors may also arise from the social environment. They may be associated, e.g., with the population density on a certain territorium (both overpopulation and isolation act as potent stressors), sexual relationships, restriction of movement, and exposure to recognized danger or to unknown situations (e.g., transport, immobilization, forced exploration of an unknown territory). Their recognition as stressors assumes a comparison with corresponding patterns stored in the memory of the organism. Some of these patterns are acquired by an individual, others are inherited and shared by all individuals of the same species ("phyletic" memory). The latter type may reflect a stress-generating situation in a distant state of phylogeny which may be rudimentary for the present species, but still remains preserved.

A well recognized biological role of the stress reaction rests in the mobilization of reserves necessary for survival during the first and second phases of exposure to noxious stimuli (cf. Fig. 2). The corresponding physiological pathways have a definite structure and the time course of the reaction is, by large, autonomic: a stressor can solely switch it on but has virtually no influence upon the course, and sometimes not even upon the intensity, of the elicited response. This conceals a potential danger that the stress reaction might be inappropriate to the stressor, and that its repetition might rather quickly lead to an exhaustion. Therefore, exposure to even a weak stressor, or its frequent repetition, may have fatal consequences for individuals with an inborn, or acquired, tendency toward an excessive stress response (Fig. 1).

Hormonal and neural pathways of stress

Physiological control

Stress, as mentioned, is a complex neurohumoral reaction which includes several hierarchically ordered processes initiated by stressors on various receptive sites (Fig. 3). Sensoric nerves conduct the signal first to the thalamus in the diencephalon and to specific neocortex regions where it evokes corresponding sensoric effects, and, at the same time, to the reticular formation of the medulla oblongata from which a further course of the stress is steered. Most of the participating reactions proceed autonomically, through sympathetic stimulation, without involvement of the higher brain centers: the tonus of both striated and smooth muscles undergo a specific control (somatomotor and visceromotor reflex, respectively), and endocrine functions of the adrenal medulla are stimulated. The visceromotor reflex, which particularly includes the tonus of vascular smooth muscle, is achieved through activation of the brain stem and results in numerous cardiovascular changes in which also the concomitantly released catecholamines and - perhaps - neurohypophyseal hormones may take

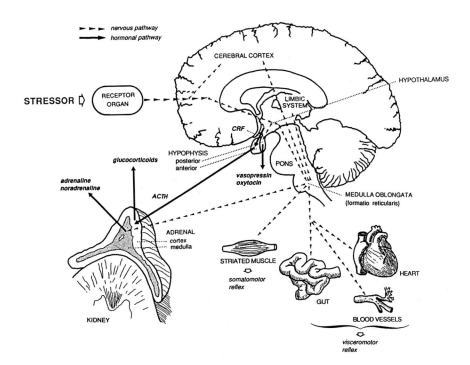

Fig. 3. *Scheme of neural and humoral steps in the alarm reaction and in the early adaptation phase of stress.*

It ought to be mentioned in passing that stress or similar reactions are phyletically very old, although the existing evidence is not sufficient to elucidate their development. They exist also in organisms with a very primitive central nervous system (like insects). Our knowledge of these primitive reaction forms is, however, still very limited.

part. The cardiac output, arterial blood pressure and heart rate are increased, the blood is newly redistributed (peripheral organs are not fully supplied), and peripheral resistance is elevated. Besides vascular muscles, other smooth muscles are also effected; so, for instance, the motility of the gut is enhanced.

Adrenal medulla

The response of the adrenal medulla to stressors is very rapid and persists during the entire alarm phase. However, all functions of the released catecholamines are not known. One of them certainly consists in enhancing the intracellular concentrations of cyclic adenosine monophosphate (cAMP) in hepatocytes (caused particularly by adrenaline), which ultimately results in an enhanced rate of glycogenolysis and a quick increase of blood glucose. It is also assumed that catecholamines may act on the limbic system and are responsible for certain emotional phenomena, like anxiety, associated with the stress.

Besides the effects on intermediary metabolism, catecholamines act as potent cardiovascular agonists with, however, different activity profiles. Whereas the α-agonists, like noradrenaline, increase the mean arterial blood pressure, decrease the pulse rate and, therefore, strongly increase peripheral resistance, the β-agonists (i.e., isoproterenol) tend toward exactly inversed effects. A massive release of catecholamines may lead to an imbalance within the cardiovascular system, with fatal consequences. In connection with various stressors, these cardiopathies have been known for a long time in various animals and also in man. It seems that the hormones of the adrenal cortex, the glucocorticoids, may exercise some kind of protective effect against an excess of catecholamines (see below).

The life-time of catecholamines in the circulation is very short (seconds to minutes), and the question arises as to the usefulness of this rapid kinetics. As far as stress regulation is concerned, one can assume that it brings about the shortest possible duration of the alarm phase which ends up with the disappearance of catecholamines. This may be an efficient energy saving mechanism in wild-living species, important particularly in undernourished animals or animals in certain states of their life cycle (for instance, during gestation).

Neurohypophysis

A decrease of the urine flow rate in cats and dogs, in response to a stress stimulus, has been known for a long time (KLISIECKI et al. 1933, RYDIN & VERNEY 1938). VERNEY (1947) ascribed it to an enhanced secretion of vasopressin which is considered to be an outcome of hypothalamic stimulation in the region of magnocellular nuclei. Since then, increased plasma levels of vasopressin were found in many cases of emotional or physical stress. Concomitant secretion of ACTH has been exceptionally observed, but in general, correlations between vasopressin and ACTH are of complex character (LUTZ et al. 1969, YATES & MARAN 1975). It is conceivable that vasopressin may influence several body functions in stress, predominantly the resorption of osmotically free water in the collecting duct of the kidney. Cardiovascular, hypophyseotropic (release of ACTH) and perhaps even some behavioral effects are likely. However, more recent experiments on rats did not fully confirm an enhancement of vasopressin secretion, not even in conditions which lead to a five-fold increase of corticosterone (ROBERTSON 1977).

Stress-dependent changes in secretion of the other neurohypopyseal hormone, oxytocin, are even more controversial. The role of this hormone in the alarm phase remains so far unclear: whereas an increase has been recorded in rats (LANG et al. 1983, GIBBS 1984), its plasma level in primates is clearly decreased (KALIN et al. 1985). It seems that its modified secretion in stress is species dependent.

Hypothalamo-pituitary-adrenocortical (HPA) axis

Energy supply activated by catecholamines can suffice for only a short period of time, owing to the rather limited glucose pools. The energy reserves which can be readily used are usually exhausted very early, already in the first phase of the stress reaction. Energy balance, but also other functions necessary for development of the resistance and for adaptation, are maintained by the processes of the HPA axis.

Neural stimuli elicited by a stressor are conducted afferently to the neurosecretory neurons in the hypothalamic parvicellular nuclei (cf. Fig. 3). These cells produce, store and eventually release the corticotropin releasing hormone (CRF), a polypeptide consisting of 41 amino acids. The site of CRF release in the hypothalamus is linked to the site of CRF action in the anterior pituitary by a specific blood vessel system, the hypothalamo-hypophysial portal circulation. After its release, CRF may reach very high concentrations in the portal blood; its concentration in the peripheral vascular bed is, however, rather low, owing to strong dilution by the peripheral blood.

In this high concentration, CRF reaches the anterior pituitary and initiates a further step of the HPA mechanism, the release of the adrenocorticotropic hormone (ACTH) from the specialized endocrine cells of the pituitary, the corticotrophs (hypophysiotropic effect of CRF). This hormone is also a polypeptide (39 amino acids) and is synthesized in corticotrophs in form of a large precursor, the pro-opiomelanocortin (POMC). Whereas the release of CRF proceeds *via* synaptic stimulation of the cell bodies of neurosecretory neurons, the ACTH release is a hormonally controlled process in which CRF receptors on the cell membrane of corticotrophs are engaged. CRF is almost certainly *not* the only factor acting hypophysiotropically within the HPA axis. As mentioned above, vasopressin may exercise analogous - or synergistic - effects by different mechanisms which are likely to be partially independent of CRF (SAFFRAN & SCHALLY 1977, BAERTSCHI & BÉNY 1982; see also contributions by GILLIES et al., BAERTSCHI et al., LUTZ-BUCHER et al., GANN & CARLSON, MCCANN et al., in BAERTSCHI & DREIFUSS 1982).

Although ACTH is virtually a glandotropic hormone, i.e., a hormone controlling secretion of another hormone from a peripheral endocrine gland, it also elicits numerous "non-tropic" responses in peripheral tissues. Some of these activities of ACTH may be highly relevant in the stress response. It is, for instance, known to act upon glycogen-glucose equilibrium, or upon lipolysis in adipocytes. Its contribution to energy supply during early phases of the stress may not be negligible. Its major role, the glandotropic effect, rests in its action on steroid-producing cells of the adrenal cortex.

In mammals, these cells are ordered in three functionally and morphologically differentiated zones (CHESTER JONES 1957, IDELMAN 1978). Their proportion and structure vary, depending on species, sex, and long-term physiological and environmental conditions of the animal (KNOBIL et al. 1954). The cells of the upper layer, zona glomerulosa, produce predominantly mineralocorticoids - steroid hormones which powerfully steer the reabsorption of sodium in the nephron. Functioning of these cells is controlled mainly by peripheral mechanisms (renin-angiotensin system: angiotensin enhances the secretion of aldosterone from glomerular cells). It seems that this mechanism is only indirectly involved in the stress response. On the other hand, the two internal zones, fasciculata and reticularis, supply steroids with manifold biological activities in stress - the glucocorticoids (cortisol, corticosterone, cortisone, 11-desoxycortisol etc.), *plus* a small amount of sexual hormones produced mostly in the zona reticularis (androgens and estrogens). The composition of the glucocorticoid spectrum displays considerable scatter not only within classes, families and species, but also among individuals of the same species. Variability can be most likely attributed to genetic background, since the expression of single enzymes involved in biosynthesis and degradation of steroid hormones may show considerable variance. It is well known that rabbits, mice and rats, for instance, synthesize only a very small amount of cortisol, if at all (BENTLEY 1982); since 17-α-hydroxylase is not expressed in a sufficient amount, corticosterone is here the dominating glucocorticoid. The macroscopic and microscopic structure of the gland also varies strongly within the vertebrate subphylum. Characteristic concentric organization of the zones around the adrenal medulla, for instance, occurs first in mammals; in birds, the aminogenic cells are localized in islets between steroidogenic cells that are mainly mineralocorticoid-producing (HARTMAN & BROWNELL 1949). In some mammalian species, zona reticularis is not distinguishable from zona fasciculata (LONG 1975), and it is doubtful whether it is present at all.

In quantitative terms, the main site of glucocorticoid synthesis is apparently zona fasciculata. The ACTH receptors on the fasciculata cells were already pharmacologically well characterized a long time ago, structure-activity relationships of steroidogenic effects of ACTH are known in many details and, also, many pathways of the cellular response (cAMP as the second messenger) have been recognized.

Glucocorticoids possess several biological functions which can be briefly listed as follows:

Metabolism of glucose. Glucocorticoids tend to maintain a constant plasma level of glucose, an effect which is of particular importance during the resistance phase of stress, or during periods of food deprivation (for instance, during hibernation or aestivation). This proceeds, first, by stimulation of gluconeogenesis, partly as a consequence of the enhanced protein catabolism (see below) and, second, by inhibition of glycolytic enzymes, which leads to the decreased glucose utilization in peripheral tissues (Fig. 4).

Protein catabolism in muscles and in extrahepatic tissues. Rate of proteolysis (particularly in muscles) and the activity of aminotransferases in the liver, are increased by glucocorticoids. The liberated amino acids are used for synthesis of carbohydrates in the liver. However, functional correlations between effects of glucocorticoids upon carbohydrate and protein metabolisms are anything but simple.

Fatty acid oxidation. Glucocorticoids act upon lipid mobilization in adipose tissues, probably by speeding up the rate of lipolysis. Free fatty acids are transported to the liver and utilized for gluconeogenesis, similar to amino acids (see above).

Antiinflammatory action. They suppress acute inflammatory responses - pain (*dolor*), swelling (*tumor*), warmth (*calor*), reddening (*rubor*), and partial immobilization (*functio laesa*) - by contraction of capillaries, prevention of hyperemia and cellular exudation, etc. They do not, however, act positively on healing processes and may even impair inflammatory conditions caused by bacteria.

Immunosuppressive effects. Glucocorticoids cause an interesting, not yet fully understood suppression of cellular immunity, manifested in several macroscopic changes like atrophy of thymus and a decrease of the lymphocyte count in the blood. Extreme effects of this kind are known from clinical situations, where glucocorticoids are employed as immunosuppressives in organ transplantation surgery. The patients are, as a result, subjected to a higher risk of malignant tumors, owing to weakened cellular defense mechanisms.

Action on the autonomic nervous system. As already mentioned, it seems likely that in certain situations glucocorticoids can suppress the activity of sympathicus and protect the organism against excessive sympathetic stimulation by catecholamines. It is well known that plasma and urinary levels of catecholamines (including their common metabolites, 4-hydroxy-3-methoxymandelic and homovanillic acids) may in certain stress-generating situations, like surgery, reach enormous values (PETRÁSEK et al. 1966). Owing to a rather slow onset of the adrenocortical response to stressors, the protective mechanisms may gain in significance first during a later stress phase. In case of a strong alarm reaction, though, this mechanism is insufficient and may not prevent a fatal "intoxication" by catecholamines. This may apply to the condition known as "capture myopathy" caused by the restraint stress (prevention of the "flight-and-fight" reaction) in wild living animals (JOHANSSON et al. 1981), or for a psychogenic stress followed by sudden exitus of a primitive tropical forest inhabitant after a magician's prophecy of death (CANNON 1957).

Experiments with adrenalectomized animals have shown that the protective effect of glucocortioids does not depend upon the ratio catecholamines to glucocorticoids. In fact, only a low concentration of cortisol is required to prevent the excessive response to catecholamines ("catecholamine intoxication"). It seems that glucocorticoids have rather a permissive effect in this respect (INGLE 1952).

Feedback control of the HPA axis

Several negative feedback pathways control secretory activities of individual organs within the HPA system (YATES et al. 1969, STOKELY & HOWARD 1972). First, CRF and ACTH secretion is regulated by the level of glucocorticoids in the peripheral blood. This happens either by their interaction with corticotrophs in the pituitary or with the neurosecretory neurons of the parvicellular neurons (long-loop feedback). Effects of cortisol on higher centers of the brain are known, but have not been narrowly defined. Secondly, ACTH exhibits a feedback control upon the hypothalamus similar to that of glucocorticoids in the so-called short-loop feedback. And thirdly, CRF can dampen its own secretion probably, again, according to its concentration in the peripheral plasma (ultrashort-loop feedback).

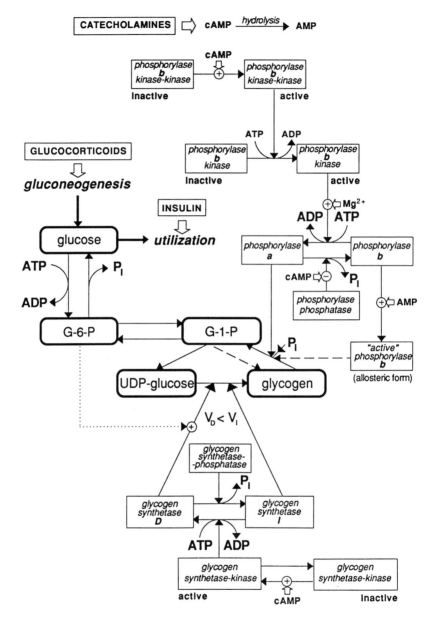

Fig. 4. Major steps of the glucose-glycogen equilibrium in the body. Right-hand side of the diagram shows salient enzymatic steps in glycogenolysis within the liver cells. Kinases exist in two functional states - inactive (dephosphorylated) and active (phosphorylated); in the first step, equilibrium between them is shifted toward the active form by cAMP. The production of cAMP in stress is elevated via activation of the cAMP-producing enzyme, the adenylate cyclase, by catecholamines, ACTH and glucagon. Possible sites of action of the other hormones relevant for a stress response are indicated. Enzymatic activity of one of the two forms of glycogen synthetase (upper part) is dependent upon the intracellular G-6-P level, the other is independent. The conversion of dependent into independent form is hormonally controlled by insulin

During a relaxed, nonstressed state of an animal, the secretion of ACTH and consequently also that of glucocorticoids show a species-dependent circadian rhythmicity, with a minimum at night and a maximum at midday. Its mechanism has not been fully recognized. The cycle seems to be determined, besides by purely neuronal processes, also by feeding times and by activity of the animal. Secretion profiles of CRF are not yet known but a pulsatile release of the hormone, analogous to the gonadotropin releasing hormone (GnRH), may come into consideration. Definitive evidence, however, is so far lacking.

Animal models of stress

Salient features of humoral stress are essentially similar among related taxonomic units. In this respect, comparative studies between humans and animals are certainly conceivable. Animal stress models of various kinds which employ small laboratory animals are in use today. Employed are manifold stressors like temperature (cold), immobilization (restraint stress), physical exercise, transport, exposure to high or low barometric pressure, water immersion, light ether anesthesia, pain, forced swimming in rats, and the like. The intensity of the elicited stress reaction is assessed mostly by changes in the adrenals (weight of the adrenals usually increases, although this is certainly not a rule), plasma levels of ACTH and/or glucocorticoids. There is no doubt that stress models are useful for physiological, pharmacological and other purposes. They are, however, problematic from the standpoint of animal protection since some degree of animal discomfort can scarcely be avoided. From this point of view, SELYE's experiments, which were so important for the precise formulation of the problem and for development of stress physiology, would probably not be permitted today in countries with highly developed animal protection laws.

A search for spontaneous stress models which would enable the treatment of animals in accord with contemporary norms is therefore of immediate interest. Farm animals may be particularly significant in this respect.

Open problems in stress research

Despite five decades of intensive research activities and thousands of scientific communications, many aspects of the stress phenomenon remain unclear. Its consequences for any individual, human or animal, as well as for human society and ecologic balance are tremendous, and can be only incompletely handled in a simple, pragmatic manner. The extraordinary position of stress among other physiological mechanisms rests in its functional vulnerability. Indeed, there exists scarcely another physiological process with a higher tendency toward excessive output, even with pathogenic consequences, than the stress response. This fact is even more astonishing since, as mentioned, it is by no means a mechanism with a *short* phyletic history. Its humoral and nervous components are virtually identical in vertebrates. On the other hand, the components associated with higher nervous activities are not only species dependent, but also differently developed between individuals of a given species. A considerable variability exists in the recognition, and in the processing, of stress-eliciting environmental and somatic conditions, owing partly to different "phyletic memories" of the species, and partly to the "intellectual" status and experience of the individual. Physiological links between the two may perhaps be the most vulnerable elements of the stress mechanism and the reason of the imperfect phylogeny of this universal phenomenon. Future basic research will, very likely, pay more attention to these aspects.

Fig. 4 (cont.): (mechanism is not fully clear) which also facilitates glucose and amino acid transport into cells (and thus indirectly the gluconeogenesis from amino acids) and participates on the control of glucose utilization by influencing the rate of kinase dephosphorylation (Krebs cycle). Compartmentisation of individual moieties (distribution between cells, tissues, etc.) is not considered in this simplified scheme. Abbreviations: G-1-P, glucose-1-phosphate; G-6-P, glucose-6-phosphate; P_i, inorganic phosphate; UDP, uridine-diphosphate; cAMP, 3',5'-cyclic adenosine monophosphate; AMP, 5'-adenosine monophosphate; V_D, V_I, maximal velocity of the glycogen synthesis for G-6-P dependent and independent enzyme forms, respectively.

Until now, methodological tools which can be called "scientific" have been applied merely to stress as a physiological process on one hand, and to its public health impacts in humans on the other. Relationships between stress and various diseases have been documented in many epidemiological statistics, prevention strategies have been at least theoretically postulated, and therapeutic measures in acute conditions associated with stress are by and large available. Attention has been paid to relations between stress and disease resistance, both in humans and in domestic animals. Less intensively investigated, if not fully neglected in the past, have been other aspects of stress, particularly those related to psychosocial stressors in animal communities. And yet, specifically this knowledge may be crucial to certain behavioral and ecologic areas, so popular nowadays, and may have practical implication for many presently discussed problems, like ethology and the keeping of domestic animals, treatment of wild animals in captivity (for instance, in zoological gardens), or protection of wild-living species. Stressors coming from social and territorial environments like isolation, expulsion from the group or territorium, hierarchy within the group (domination - subordination), sexual interactions (interesting pathophysiological observations on chickens have been reported by RATCLIFFE & SNYDER 1964) may be very potent, are physiologically less expedient and reflect a real danger for animals, or even humans. In general, however, wild-living animals are more sensitive to psychosocial stressors than are domesticated species; this can be inferred, e.g., from the comparison of adrenal characteristics in wild *versus* laboratory rats (EBINGER 1972).

The available evidence seems to indicate that the main function of stress is to secure the survival of an organism during emergencies, and there is no particular reason to doubt this role. However, one fact is certainly not in keeping with this notion: the suppressive effect of stress on the immune defense. This is an obvious disadvantage for the organism, particularly during a long lasting adaptation phase in which, due to an enhanced glucocorticoid secretion, the organism is exposed both to pathogenic agents from the environment and to a "self-destruction" by malignant neoplastic processes (LEWIS & PHILLIPS 1979). A question arises as to whether stress does not have an additional function which pertain to the social unit: to "label" individuals burdened with the clear-cut physiological disadvantage of recovering only slowly from stress, and to enhance the probability of their elimination. The elimination process would lead to a natural selection of individuals that can better cope with environmental stressors. In the long run, it may become interesting - and important - to know the genomic relationships between the "stress-linked" genes and the genes determining behavioral features like aggressivity, learning ability or social dominance within the animal group, and to recognize these genes, if they exist.

Sources of literature

As this has been, in effect, merely an abbreviated consideration of the problem of stress in animals, most references to the original articles have been omitted. Only less well known, or historically interesting, contributions have been cited. The presentation itself rests upon the fundamental ideas of H. SELYE summarized in his numerous books and publications, particularly in "Stress" (1950), "The Stress of Life" (1956) and "*In Vivo* - The Case of Supramolecular Biology" (1967). The source of some notions employed here is J. CHARVÀT (1897- 1984), professor of internal medicine, and former chairman of the 3rd Internal Clinic and Laboratory of Endocrinology, Charles University, Prague (Czechoslovakia). His book "Life, Adaptation and Stress", available unfortunately only in the Czech original ("Zivot, adaptace a stress", Avicenum, Prague 1970) contains many remarkable observations and comments on the stress phenomenon.

A comprehensive treatment of stress physiology in farm animals is presented in the handbook "Stress Physiology in Livestock" edited by M.K. YOUSEF (CRC Press, Inc., Boca Raton, FL, 1985; Vol. 1: Basic Principles, Vol. 2: Ungulates, Vol. 3: Poultry).

References

BAERTSCHI, A.J.; BÉNY, J.L. (1982). Central control of ACTH secretion: role of vasopressinergic pathways and of the hypothalamohypophysial circulation. Front. Horm. Res. **10**, 126-140.

BAERTSCHI, A.J.; DREIFUSS, J.J. (Editors) (1982). Neuroendocrinology of Vasopressin, Corticoliberin and Opiomelanocortins. Academic Press, London.

BENTLEY, P. (1982). Comparative Endocrinology of Vertebrates, 2nd Edition. Cambridge University Press, Cambridge, U.K., P. 73.

CANNON, W.B. (1957): Woodoo death. Psychosom. Med. **19**, 182-190.

CHESTER JONES, I. (1957): The Adrenal Cortex. Cambridge University Press, Cambridge, U.K.

EBINGER, P. (1972): Vergleichend-quantitative Untersuchungen an Wild- und Laborratten. Z. Tierzchtg. Züchtungsbiol. **89**, 34-57.

GIBBS, D.M. (1984): Dissociation of oxytocin, vasopressin and corticotropin secretion during different types of stress. Life Sci. **35**, 487-491.

HAKEN, H. (1978): Synergetics - An Introduction. Springer-Verlag, Berlin.

HAKEN, H. (Editor) (1982): Evolution of Order and Chaos in Physics, Chemistry, and Biology. Springer-Verlag, Berlin.

HARTMAN, F.A.; BROWNELL, K.A. (1949): The Adrenal Gland. Lea and Febiger, Philadelphia.

IDELMAN, S. (1978): The structure of the mammalian adrenal cortex. In: General, Comparative and Clinical Endocrinology of the Adrenal Cortex, Vol. 2 (Editors: CHESTER-JONES, I.; HENDERSON, I.W.). Academic Press, London. Pp. 1-199.

INGLE, D.J. (1952): The role of the adrenal cortex in homeostasis. J. Endocrinol. **8**, xxiii-xxxvii.

JOHANSSON, G.; JÖNSSON, L.; THORÉN-TOLLING, K.; HAGGENDAL, J. (1981): Porcine stress syndrome - a general response to restraint stress. In: Porcine Stress and Meat Quality (Eds.: FROYSTEIN, T.; SLINDE, E.; STANDAL, N.). Agricultural Food Research Society, Norway. Pp. 32-41.

KALIN, N.H.; GIBBS, D.M.; BARKSDALE, C.M.; SHELTON, S.E.; CARNES, M. (1985): Behavioral stress decreases plasma oxytocin concentrations in primates. Life Sci. **36**, 1275-1280.

KLISIECKI, A.; PICKFORD, M.; ROTHSCHILD, P.; VERNEY, E.B. (1933): The absorption and secretion of water by the mammals. Proc. Roy. Soc. London B **112**, 496-547.

KNOBIL, E.; MORSE, A.; HOFMANN, F.G.; GREEP, R.O. (1954): A histologic and histochemical study of hypophyseal-adrenal cortical relationships in the rhesus monkey. Acta Endocrinol. **17**, 229-238.

LANG, R.E.; HEIL, J.W.E.; GANTEN, D.; HERMANN, K.; UNGER, T.; RASCHER, W. (1983): Oxytocin unlike vasopressin is a stress hormone in the rat. Neuroendocrinology **37**, 314-316.

LEWIS, M.G.; PHILLIPS, T.M. (1979): The possible effects of emotional stress on cancer mediated through the immune system. In: Cancer, Stress, and Death (Editors: TACHE, J.; SELYE, H.; DAY, S.B.). Plenum Medical Book Comp., New York. Pp. 21-27.

LONG, J.A. (1975): Zonation of the mammalian adrenal cortex. In: Handbook of Physiology, Section 7: Endocrinology (Section Editors: GREEP R.O., ASTWOOD E.B.), Vol. 6: Adrenal Gland (Volume Editors: BLASCHKO, H.; SAYERS, G.; SMITH, A.D.). American Physiological Society, Washington, D.C. Pp. 13-24.

LUTZ, B.; KOCH, B.; MIALHE, C. (1969): Libération des hormones antidiurétique et corticotrope au cours de différents types d'agression chez le rat. Horm. Metab. Res. **1**, 213-217.

PETRÁSEK, J.; DUBOVSKY, J.; CHARVÁT, J. (1966): Clinical evaluation of catecholamine excretion. Endokrinologie **50**, 308-316.

RATCLIFFE, H.L.; SNYDER, R.L. (1964): Myocardial infarction: a response to social interaction among chickens. Science **144**, 425-426.

ROBERTSON, G.L. (1977): The regulation of vasopressin function in health and disease. Rec. Prog. Horm. Res. **33**, 333-385.

RYDIN, H.; VERNEY, E.B. (1938): The inhibition of water-diuresis by emotional stress and by muscular exercise. Quart. J. Exp. Physiol. **27**, 343-374.

SAFFRAN, M.; SCHALLY, A.V. (1977): The status of the corticotropin releasing factor (CRF). Neuroendocrinology **24**, 359-375.

SELYE, H. (1936): A syndrome produced by diverse nocuous agents. Nature (London) **138**, 32.

SELYE, H. (1950): Stress. Acta Inc. Med. Publ., Montreal.

SELYE, H. (1956): The Stress of Life. Longmans, Green & Co., London.

SELYE, H. (1967): In Vivo - The Case of Supramolecular Biology. Liveright Publishing Corporation, New York.

SELYE, H.; PENTZ, E.I. (1943): Pathogenetical correlations between periarteritis nodosa, renal hypertension and rheumatic lesions. Can. J. Med. Assoc. **49**, 264-272.

STOKELY, E.M.; HOWARD, L.L. (1972): Analog computer model for the ACTH- glucocorticoid system. IEEE Transactions on Bio-Medical Engineering BME **19**, 3-20.

VERNEY, E.B. (1947): The antidiuretic hormone and the factors which determine its release. Proc. Roy. Soc. London B **135**, 25-106.

YATES, F.E.; BRENNAN, R.D.; URQUHART, J. (1969): Adrenal glucocorticoid control system. Feder. Proc. **28**, 71-83.
YATES, F.E.; MARAN, J.W. (1975): Stimulation and inhibition of adrenocorticotropin release. In Handbook of Physiology, Section 7: Endocrinology (Section Editors: GREEP, R.O.; ASTWOOD, E.B.), Vol. **4**, Part 2: The Pituitary Gland and its Neuroendocrine Control (Volume Editors: KNOBIL, E.; SAWYER, W.H.): American Physiological Society, Washington, D.C. Pp. 367-404.

Author's address:
V. Pliška, Institut für Nutztierwissenschaften, ETH Zürich, CH-8092 Zürich, Switzerland.

Stress in domestic pigs

J. J. HARI and E. LANG

Definition of stress

Despite environmental changes and varying metabolic needs, the body regulates its interior conditions within fixed limits. Stress is considered to be a chain of changes exceeding the body's coping capacity. Difficulty in reacting to such demands leads to numerous disorders in endocrine, metabolic or immune systems, and in behavior (SELYE 1976).

Historically, the concept of stress has always emphasized changes in the endocrine system. It is now becoming more and more evident that psychological factors contribute strongly to the stress syndrome (DANTZER & MORMÈDE 1983, DANTZER & MORMÈDE 1985, LEVINE 1985). Therefore, a definition of stress is increasingly difficult. DANTZER and MORMÈDE (1985) believe that the concept of stress has little explanatory value. For that very reason, it is always indispensable to define the term "stress" in order to avoid confusion. H. SELYE (1976), who mainly contributed to the development of the stress concept, defines stress as the nonspecific response of the body to any demand.

A question arises as to what can be considered as "nonspecific". As endocrine and metabolic events are regulated by a large variety of positive and negative feedback mechanisms, and are also closely linked to each other, it is difficult to decide if a particular response is specific or nonspecific. Increases in plasma levels of ACTH, cortisol and catecholamines are largely accepted as indicators of short-term stress. As indicators of long-term stress, enlargement of the adrenals is commonly accepted. These changes always occur during situations which can be considered as stressful. An event that elicits such changes is called a stressor.

Pigs are known to be very sensitive to stress in general. Remarkably, this sensitivity involves a genetic component. With increasing breeding intensity certain strains of pigs became very sensitive to even moderate stressors. An increase in losses due to sudden death of pigs during transport, or other manipulations has often been observed. These losses have been correlated with decreased meat quality, i.e., a higher incidence of pale, soft, and exudative (PSE) meat (for review see MARPLE & CASSENS 1973b). In the early 70's EIKELENBOOM & MINKEMA (1974) found a correlation between the occurrence of malignant hyperthermia during halothane anaesthesia and the disorders mentioned above. The pigs that developed malignant hyperthermia (referred to as Hal+) were termed "stress sensitive". The Hal+ pigs have been systematically excluded from further breeding. It should be noted that this selection greatly reduced losses due to bad meat quality and losses during stressful situations. We believe that the exclusion of Hal+ pigs helped to solve many problems in pig breeding but that the problem of stress sensitivity itself has still not been solved (HARI et al. 1986).

Reaction to stressors

The nonspecific stress reaction is outlined in detail in the article by PLIŠKA in this monograph. Fig. 3 of that article shows the main organs and hormones involved in the nonspecific defense reaction. The following paragraphs briefly add some details, which are important, in our context.

VALE and coworkers (1981) isolated the hypophyseal corticotropin releasing factor (CRF) of the sheep and determined its amino acid sequence. The CRF of the pig differs in 8 amino acid residues (PATTY et al. 1986). CRF stimulates, *in vitro* and *in vivo*, the release of adrenocorticotropic hormone (ACTH) from the pituitary (ANTONI 1986, RIVIER et al. 1982a, RIVIER & VALE 1983). Several hormones modify the effect of CRF or act, themselves, as releasing factors (AIZAWA et al. 1982, ANTONI 1986, AXELROD & REISINE 1984, KELLER-WOOD & DALLMAN 1984, RIVIER et al. 1984a). CRF is the most important, although not the single releasing factor, of ACTH. The passive immunoneutralization of CRF inhibits ACTH secretion during stress (RIVIER et al. 1982b, CONTE-DEVOLX et al. 1983); a CRF antagonist has similar effects (RIVIER et al. 1984b). Glucocorticoids and ACTH reduce chemically and immunhistologically detectable CRF in the hypothalamus, indicating a negative feedback on CRF production (ANTONI 1986, HAUGER et al. 1987, CHAPPELL et al. 1986). CRF has different central nervous effects (ZADINA et al. 1986).

ACTH is cleaved from a larger precursor molecule, the proopiomelanocortin (POMC) (MAINS et al. 1977, MARKS 1978, CELIO et al. 1980), and has a lipolytic effect in adipose tissue (RICHTER et al. 1987, RICHTER & SCHWANDT 1987, see also ARMARIO et al. 1986).

The corticosteroids endow the organism with the capacity to resist stressful situations and noxious stimuli (MUNCK et al. 1984). In our context, the positive effects on gluconeogenesis, fat mobilization, and the negative effect on insulin action are of particular interest. The exact regulating role of cortisol still remains to be clarified in detail. In general, cortisol is a catabolic hormone (JONES & HENDERSON 1980).

Genetically determined stress reaction?

As mentioned above, a correlation exists between the reaction to stressful situations and meat quality or, in broader terms, to energy metabolism. Pigs selected for low body fat content became increasingly sensitive to various stimuli (for review see MARPLE & CASSENS 1973b). JUDGE et al. (1968) found lower glucocorticoid levels in stress-sensitive pigs. They concluded that these pigs suffer from an adrenal insufficiency. MARPLE et al. (1969) induced such an insufficiency in pigs by treating them with prednisolone. The pigs became, indeed, stress-sensitive but the treatment was without any negative influence on meat quality. In consequence, the hypothesis of adrenal insufficiency has to be rejected (ROGDAKIS et al. 1975).

During the early 70's several authors carefully investigated the pituitary-adrenal axis. MARPLE et al. (1972a) showed that stress-sensitive pigs have higher plasma levels of ACTH than stress resistant pigs, whereas the cortisol levels did not differ. Further studies reveald similar results (MARPLE et al. 1972b, MARPLE & CASSENS 1973b). In an infusion study, MARPLE & CASSENS (1973a) found an increased metabolic clearance rate of cortisol in stress sensitive pigs. This accounts for the differences found in ACTH and cortisol plasma levels. Unfortunately, these interesting results did not encourage further research.

Circulating cortisol is partially bound to cortisol binding globulin (CBG). This protein influences the biological activity of cortisol but the blood plasma levels of this protein vary less than do cortisol plasma levels. KATTESH et al. (1980) considered it as a suitable indicator of stress sensitivity in pigs. Other authors had already obtained similar results earlier (MARPLE et al. 1974a, ABERLE et al. 1976), but they only found differences between races of pigs with different degrees of stress sensitivity; within the same race, there was no difference in CBG levels between stress-sensitive and stress resistant pigs. This parameter is therefore generally considered to be a rather poor predictor of stress sensitivity.

Another criterium of stress sensitivity, the so-called "halothane reaction" (see above), was introduced by EIKELENBOOM & MINKEMA (1974). This halothane test became soon a well accepted predictor of stress sensitivity in pigs. Shortly after inhalation of halothane, certain genetically determined pigs develop a malignant hyperthermia (ARAKI et al. 1985, EIKELENBOOM & MINKEMA 1974, VON FABER et al. 1983) as a consequence of a disturbed intracellular calcium metabolism (LUCKE et al. 1976, OHNISHI et al. 1986, OHNISHI 1987). This leads to a tetanic contraction of striated muscles and a stimulation of glycogenolysis. Because of the rapid development of the reaction, glycogen is broken down under an oxygen deficit and large quantities of lactate are formed. Lactates act as a stressor, stimulate catecholamine release, and in consequence, further enhance glycogenolysis. In addition, the strong muscle contractions cause hyperthermia. If the inhalation of halothane is not ceased, the animal dies within a few minutes (VON FABER et al. 1983, see also HALL et al. 1976a, HALL et al. 1976b, HALL et al. 1977, HALL et al. 1980, LUCKE et al. 1976, LUCKE et al. 1978, LISTER et al. 1976).

It is assumed that similar reactions proceed in pigs *post mortem* in the slaughterhouse. Lactate is accumulated in the muscle, and the pH in the muscle increases. Its measurement with inserted pH electrodes does, however, not reflect the actual pH of the muscle but serves as an indicator of bad meat quality. The Hal+ genotype is inherited in an autosomal recessive way (MABRY et al. 1981,

MABRY et al. 1983, REIK et al. 1983) and can slightly be influenced by other genes. It is quite often linked to bad meat quality and increased stress sensitivity (GERWIG et al. 1979, VÖGELI et al. 1984, WEBB 1981).

Investigations on obese and non-obese pigs

On an industrial scale, pigs are mainly bred for high body weight gain (to reduce costs of maintenance), and for low fat content (to meet consumer's demands). Several pig races and strains are commonly bred in the western countries, and attempts to correlate metabolic parameters with different selection criteria are therefore problematic. Therefore, pig strains of a single race that are selected in two opposite directions, according to defined selection criteria, are particularly useful to examine the physiological mechanisms underlying the selection. Two such selection studies, which include a variety of measurements of physiological parameters, have been independently performed so far. The first experiments were conducted at the University of Hohenheim, Stuttgart, FRG (ROGDAKIS 1982, MÜLLER & ROGDAKIS 1985). The selection was based on the activity of NADPH-generating enzymes in adipose tissue: one line was selected for high activity, the other one for low activity; another line was selected for low back fat thickness (measured by ultrasound), and one line served as control. The investigators measured a large number of metabolic parameters such as insulin, glucose, cortisol and others.

The second series of experiments were conducted at the ETH in Zürich, Switzerland (VÖGELI & GERWIG 1979, VÖGELI et al. 1982, VÖGELI et al. 1984, WÄFLER et al. 1982). One line was selected for high back fat thickness and low daily body weight gain, the other line was selected for low back fat thickness and high daily body weight gain. In both experiments the lines differed markedly in body fat content, growth rate and stress sensitivity after 8 generations of breeding. Again, a number of physiological parameters was assessed.

These experiments revealed two important points. First, stress sensitivity involves a genetic component and second, stress sensitivity is closely linked to energy metabolism, and consequently, to the degree of obesity. The term "obesity" does not necessarily mean that pigs are obese in the clinical sense of the word, but it implies differences in body fat content.

Recent findings

In both the Zürich and the Hohenheim studies, the breeding lines differed markedly in back fat thickness (estimated body fat content) and stress sensitivity. In the Zürich experiment, the difference in back fat thickness was the consequence of the selection process. In the Hohenheim experiments, the breeding progress was based on the selection of different enzyme activities. This indicates that the chosen NADPH-generating enzymes are among the central components involved in genetic obesity in pigs. The data from these experiments confirmed earlier observations by the same authors (ROGDAKIS et al. 1979, STRUTZ & ROGDAKIS 1979). In the Hohenheim experiments, there was one surprising point: the total enzyme activity did not increase in the line selected for high enzyme activity, despite an increase in adipose tissue mass (MÜLLER & ROGDAKIS 1985). Some additional factors probably were responsible for this finding.

In both studies, the selection did not include any parameter of stress sensitivity. Nevertheless, the lines showed differences in that respect. The selection procedure obviously favored pigs with altered endocrine activity. One could also hypothesize that animals with unfavorable metabolic balances resulted from the breeding strategy. Metabolic disturbances would then act as possible stressors and the animals would be chronically stressed and show signs of increased stress sensitivity. This hypothesis is, however, very unlikely (see below). The following changes in endocrine activity have been partially characterized.

Glucose tolerance. In both studies, a glucose tolerance test was performed. The glucose and insulin levels following administration of intravenous glucose injection showed that pigs with high body fat content were insulin resistant (Fig. 1; KALBITZ 1986). The baseline values for insulin were also higher in these pigs. KALBITZ (1986) measured insulin receptor and post-binding events. They found equal

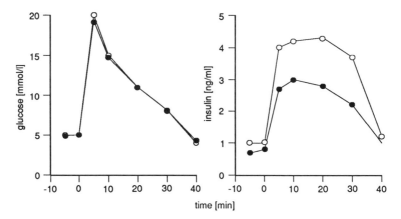

Fig. 1. Glucose tolerance in obese and non-obese pigs. Pigs originated from the 5th and 7th generation of the Zürich experiments. They weighed approximately 60 kg. Glucose was infused i.v. from 0 to 5 minutes in catheterized pigs and blood was withdrawn through the same catheter. The left-hand panel shows the glucose levels, the right-hand panel shows the insulin levels. Open circles: obese pigs; closed circles: nonobese pigs. Results are arithmetic means of 31 animals in each group. From SCHNEEBELI (1987), modified.

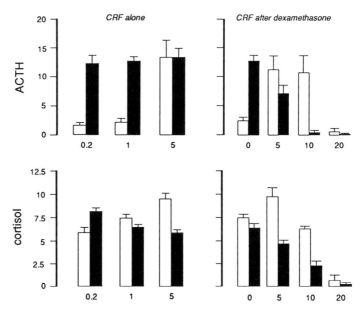

Fig. 2. ACTH and cortisol release after an injection of CRF in obese (white bars) and non-obese pigs (black bars). Pigs originated from the 8th generation of the Zürich experiments. They weighed approximately 30 kg. CRF was i.v. injected in catheterized (TAKAHASHI 1986) pigs. Blood was withdrawn through the catheter during a period of 180 minutes. The ordinares show the total (integrated) release of the hormones in ng over the whole experimental period of 180 min, per ml plasma (area under the stimulation curve). Left hand panels: CRF alone; CRF doses in µg/kg b.w. are indicated below the bars. Right-hand panels: pretreatment with dexamethasone (doses in µg/kg b.w. below the bars) 4 hours prior to an injection of 1 µg of CRF. The upper panels show the total ACTH release; the lower panels show the corresponding cortisol release. Data represent the arithmetic means and standard errors of mean (10 animals in each group.)

affinity and concentrations of the receptor in both groups. The non-obese pigs had a higher insulin sensitivity. In the Zürich experiments, WÄFLER and coworkers (1982) also measured glucagon levels. The obese pigs had lower concentrations of glucagon than their non-obese counterparts.

Growth hormone. Both research groups found no differences in baseline growth hormone levels (MÜLLER & ROGDAKIS 1985, WÄFLER et al. 1982).

Glucocorticoids and ACTH. In both studies, baseline cortisol levels were higher in the obese group. Stimulated cortisol levels differed between obese and non-obese pigs as well. In the Zürich experiments we measured the ACTH and cortisol output after a bolus injection of CRF (Fig. 2). After the injection, the obese pigs released ACTH in a dose dependent manner, whereas the non-obese pigs showed a maximal release at all three doses of CRF. Dexamethasone showed an unexpected inhibitory effect on ACTH release in the non-obese group (Fig. 2). This indicated an altered feedback mechanism. In the same group of pigs, we estimated the plasma half life of cortisol. The half life in the obese group was 11.4 minutes and, in the non-obese group, 6.9 minutes. The faster disappearance rate of cortisol in the non-obese group led to an altered feed back mechanism on ACTH output and, in consequence, to the observed changes in ACTH release (HARI 1988).

A bolus injection of insulin, a commonly used stressor (FISH et al. 1986), showed similar results: the non-obese pigs released more ACTH than the obese pigs (Fig. 3). The histological examination of the adrenals and the pituitaries confirmed these results (PLIŠKA et al., in preparation): the adrenals of the non-obese animals showed a small enlargement of the zona fasciculata and of the total adrenal weight. These measurements suggested the effect of a weak stressor. In contrast, the ACTH-producing cells of the pituitaries of the two lines differed markedly: these cells were enlarged in the non-obese line, suggesting the effect of a strong stressor. The small changes in adrenals and great changes in pituitaries can be explained by different cortisol half-life as a consequence of different feedback mechanisms.

Catecholamines. The lipolytic activity of catecholamines has been characterized in both lines with different methods. It seems that the lipolytic activity is lower in obese pigs. This can be explained by their higher insulin levels, because insulin is the strongest antilipolytic hormone. Many other results have been obtained: the ratio of adrenaline and noradrenaline is different in obese and non-obese

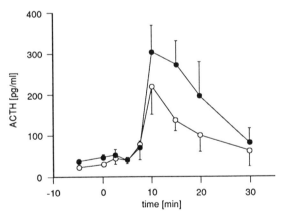

Fig. 3. ACTH release after an injection of insulin. The figure shows the ACTH release after an injection of insulin (arithmetic means and standard errors of mean). Pigs are the same as those described in fig. 2. Insulin (2.15 µmol/kg) was injected i.v. in catheterized pigs at the time zero. Open circles: obese pigs; closed circles: non-obese pigs.

animals, the obese animals have a higher concentration of β-receptors in adipose tissue, etc. (MÜLLER & ROGDAKIS 1985, SCHNEEBELI 1987). These results are quite often confusing and difficult to interpret.

Other biochemical parameters. Some other parameters, such as triiodothyronine, testosterone, and plasma urea, were measured in both studies. Most of the measurements revealed differences between obese and non-obese pigs. This suggests that a variety of metabolic pathways have been changed during the selection process. This is probably only a consequence and not a direct effect of the selection process, because the lines already differed in both experiments after few generations (two to four) in many parameters, and it is unlikely that so many metabolic pathways had been changed concomitantly.

A role for cortisol in obesity?

It is very likely that only few genes are involved in the above mentioned selection experiments with pigs. A single gene can already account for great changes in a large number of metabolic processes. Two examples of such single gene mutations in another species may illustrate this idea: the homozygous fa/fa Zucker rats become obese and insulin resistant, whereas heterozygous animals are normal (BRAY & YORK 1979). The gene itself has not yet been identified. The Brattleboro rat, as the second example, has a mutated vasopressin gene (SCHMALE & RICHTER 1984). This single point mutation has a profound effect upon water and salt metabolism.

We hypothesize that many of the present selection criteria favor animals with a short half-life (parallels high clearance rate) of cortisol. Fig. 4 summarizes this hypothesis. The following points may explain the its basis.

i) Glucocorticoids are known to cause insulin resistance (NATIONAL DIABETES DATA GROUP 1979). Fig. 5 shows an example of such an experimentally induced insulin resistance. Rats were injected daily during two weeks with corticosterone. Two days after this treatment the rats underwent a glucose tolerance test. The insulin data show that the treated rats are indeed insulin resistant. Pigs with a long half-life of cortisol would be slightly insulin resistant compared to pigs with a shorter half life of the hormone. The insulin resistance is confirmed in several publications (WÄFLER et al. 1982; ENSINGER et al. 1979).

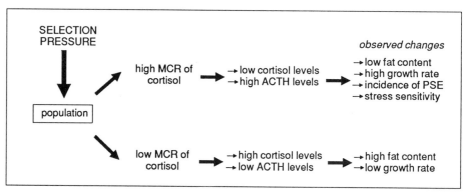

Fig. 4. A hypothesis on the development for stress-sensitivity and low body fat content in domestic pigs. Domestic pigs had been selected for breeding according to their low body fat content and their high growth rate. This selection process favored pigs with a high clearance rate of cortisol, which led to the desired (and also some undesired!) characteristics. MCR, metabolic clearance rate; PSE, "pale - soft - exudative" muscles.

ii) The plasma half-life of cortisol in pigs is lower than in other species (HARI 1988). This may be a consequence of selection during recent decades. Even so-called stress resistant pigs in recent experiments were once selected for high growth rate and low body fat content. A direct comparison of "modern" pigs with wild boars would clarify this point.

iii) From obese rodents it is known that adrenalectomy prevents obesity (DUBUC & WILDEN 1986, FREEDMAN et al. 1986), and that replacement of corticosterone restores the development of obesity (CASTONGUAY et al. 1986). This demonstrates the importance of glucocorticoids in fat metabolism (see also AMATRUDA et al. 1983, DIAMANT & SHAFRIR 1975, KING & SMITH 1985).

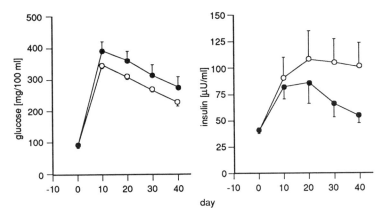

Fig. 5. Corticosterone-induced insulin resistance in rats. Normal rats were injected s.c. daily, for two weeks with corticosterone diluted in sesame oil. Two days after this treatment a glucose tolerance test was performed. 1 g/kg glucose was injected i.v. at time zero. The left-hand panel shows the plasma glucose, the right-hand panel the plasma insulin levels. Open circles: corticosterone treatment; closed circles: controls (6 animals per group, arithmetic means and standard errors of mean; ref. see text).

iv) Insulin-resistant animals show elevated baseline insulin levels. Insulin is known to stimulate, for example, glucosephosphate-dehydrogenase (SALATI et al. 1988). Therefore, elevated activities of this enzyme might be a consequence, but not the cause, of insulin resistance in the Hohenheim experiments.

v) MUNCK et al. (1984) reconsidered the role of glucocorticoids in stress and proposed that glucocorticoids protect the body from overshooting of its normal defense mechanism against stressors. Thus, pigs with a higher metabolic clearance rate of cortisol become more stress-sensitive, because their normal defense reactions are exceeded. Therefore, they also show signs of chronic stress.

Conclusions

Pigs are useful models in biomedicine. Their high body weight, however, limits their use in *in vivo* experiments. The close relationship between stress sensitivity and fat metabolism can be of particular interest. This makes the pig particularly useful model in the field of obesity, diabetes type II (non-insulin dependent diabetes), atherosclerosis, and other related disorders. Some examples may illustrate the potential use of pigs in biomedicine.

i) A genetic component is clearly involved in human diabetes type II (ELBEIN et al. 1988, O'RALILLY et al. 1988). Under the same risk factors, genetically determined insulin resistant pigs may develop similar symptoms.

ii) In humans, insulin resistance is correlated with hypertension and atherosclerosis (REAVEN & HOFFMAN 1987, REAVEN & HOFFMAN 1988, LUCUS et al. 1985). This may also be the case in pigs (see also RAPACZ & RAPACZ-HASLER, this monograph).

iii) Obese pigs have higher glucagon levels than non-obese pigs similar observations having been made in humans (REAVEN et al. 1987).

iv) Certain pigs develop a malignant hyperthermia during halothane anaesthesia. This disease also resembles the one known in humans (ARAKI et al. 1985, OHNISHI et al. 1986, OHNISHI 1987).

References

ABERLE, E. D.; RIGGS, B. L.; ALLISTON, C. W.; WILSON, S. P. (1976): Effects of thermal stress, breed and stress susceptibility on corticosteroid binding globulin in swine. J. Anim. Sci. **43**, 816-820.

Aizawa, T.; Yasuda, N.; Greer, M. A.; Sawyer, W. H. (1982): *In vivo* adrenocorticotropin- releasing activity of neurohypophyseal hormones and their analogs. Endocrinology **110**, 98-104.

Amatruda, J. M.; Danahy, S. A.; Chang, C. L. (1983): The effects of glucocorticoids in insulin-stimulated lipogenesis in primary cultures of rat hepatocytes. Biochem. J. **212**, 135-141.

Antoni, F. A. (1986): Hypothalamic control of adrenocorticotropin secretion: advances since the discovery of 41-residue corticotropin-releasing factor. Endocrinol. Rev. **7**, 351-378.

Antoni, F.A.; Holmes, M.C.; Jones, M.T. (1983): Oxytocin as well as vasopressin potentiate ovine CRF *in vitro*. Peptide **4**, 411-415.

Araki, M.; Takagi, A.; Fujita, T.; Matsubara, T. (1985): Porcine malignant hyperthermia: caffeine contracture of single skinned muscle fibers. Biomed. Res. **6**, 73-78.

Armario, A.; Campmany, L.; Hidalgo, J. (1986): Metabolic effects of chronic ACTH administration, interaction with response to stress. Pharmacology **33**, 235-240.

Axelrod, J.; Reisine, T. D. (1984): Stress hormones: their interaction and regulation. Science **224**, 452-459.

Bray, G. A.; York, D. A. (1979): Hypothalamic and genetic obesity in experimental animals: an autonomic and endocrine hypothesis. Physiol. Rev. **59**, 719-809.

Castonguay, T. W.; Dallamann, M. F.; Stern, J. S. (1986): Some metabolic and behavioral effects of adrenalectomy on obese Zucker rats. Amer. J. Physiol. **251**, R 923-933.

Celio, M. R.; Pasi, A.; Bürgisser, E.; Buetti, G.; Höllt, V.; Gramsch, C. (1980): Proopiocortin fragments in normal human adult pituitary. Distribution and ultrastructural characterization of immunoreactive cells. Acta Endocrinol. **95**, 27-40.

Chappell, B.; Smith, M. A.; Kilts, C. D.; Bissette, G.; Ritchie, J.; Anderson, C.; Nemeroff, C. B. (1986): Alterations in corticotropin-releasing factor-like immunoreactivity in discrete rat brain regions after acute and chronic stress. J. Neurosci. **6**, 2908-2914.

Conte-Devolx, B.; Rey, M.; Boudouresque, F.; Giraud, P.; Castanas, D.; Millet, Y.; Codaccioni, J. L.; Oliver, C. (1983): Effect of 41-CRF antiserum on the secretion of ACTH, beta-endorphin and alpha-MSH in the rat. Peptides **4**, 301-304.

Dantzer, R.; Mormède, P. (1983): Stress in farm animals: a need for reevaluation. J. Anim. Sci. **57**, 6-18.

Dantzer, R.; Mormède, P. (1985): Stress in domestic animals: a psychoneuroendocrine approach. In: Animal Stress. (Editor: Moberg, G. P.). Amer. Physiol. Soc., Bethesda, Maryland, USA. Pp. 81-95.

Diamant, S.; Shafrir, E. (1975): Modulation of the activity of lipogenesis by glucocorticoids. Eur. J. Biochem. **53**, 541-546.

Dubuc, U.; Wilden, N. J. (1986): Adrenalectomy reduces but does not reverse obesity in ob/ob mice. Int. J. Obesity **10**, 91-98.

Eikelenboom, G.; Minkema, D. (1974): Prediction of pale, soft, exudative muscle with non-lethal test for the halothane-induced porcine malignant hyperthermia syndrome. Tjdschr. Diergeneesk. **99**, 421-426.

Elbein, S. C.; Ward, W. K.; Beard, J. C.; Permutt, M. A. (1988): Familial NIDDM: Molecular-genetic analysis and assessment of insulin action and pancreatic B- cell function. Diabetes **37**, 377-382.

Ensinger, U.; Rogdakis, E.; von Faber, H. (1979): Glukosetoleranz und Insulinsekretion bei Piétrains und Edelschweinen. Z. Tierphys. Tierernährung Futtermittelkunde **41**, 301-309.

Faber, H.,v.; Rogdakis, E.; Müller, E. (1983): Die Rolle von Hormonen bei der Entstehung von PSE-Fleisch - Eine Uebersicht. Züchtungskunde **55**, 337-347.

Fish, H. R.; Chernow, B.; O'Brian, J. T. (1986): Endocrine and neurophysiologic responses of the pituitary to insulin-induced hypoglycemia: a review. Metabolism **35**, 763-780.

Freedman, M. R.; Stern, J. S.; Reaven, G. M.; Mondon, C. E. (1986): Effect of adrenalectomy on *in vivo* glucose metabolism in insulin resistant Zucker obese rats. Horm. Metab. Res. **18**, 296-298.

Gerwig, C.; Vogeli, P.; Schwörer, D. (1979): Halothane sensitivity in a positive and a negative selection line. Acta Agric. Scand. Suppl. **21**, 441-450.

Hall, G. M.; Bendall, J. R.; Lucke, J. N.; Lister, D. (1976a): Porcine malignant hyperthermia. 2. Heat production. Brit. J. Anaesth. **48**, 305-308.

Hall, G. M.; Lucke, J. N.; Lister, D. (1976b): Porcine malignant hyperthermia. 4. Neuromuscular blockade. Brit. J. Anaesth. **48**, 1135-1140.

Hall, G. M.; Lucke, J. N.; Lister, D. (1977): Porcine malignant hyperthermia. 5. Fatal hyperthermia in the Piétrain pig, associated with the infusion of alpha- adrenergic agonists. Brit. J. Anaesth. **49**, 855-863.

Hall, G. M.; Lucke, J. N.; Lovell, R.; Lister, D. (1980): Porcine malignant hyperthermia. 7. Hepatic metabolism. Brit. J. Anaesth. **52**, 11-17.

Hari, J. J. (1988): Hypophysen-Nebennieren Achse bei fetten, stressresistenten und mageren, stressempfindlichen Schweinen. Thesis No. 8457, ETH Zürich, Switzerland.

Hari, J.; Heiniger, J.; Pliška, V. (1986): Das endokrine System im Stresssyndrom beim Schwein. Bulletin ETH Zürich **198**, 17-18.

HAUGER, R. L.; MILLAN, M. A.; CATT, K. J.; AUILERA, G. (1987): Differential regulation of brain and pituitary corticotropin-releasing factor receptors by corticosterone. Endocrinology 120, 1527-1533.
JONES, I. C.; HENDERSON, I. W. (Editors) (1980): General comparative and clinical endocrinology of the adrenal cortex. Academic Press, London.
JUDGE, M. D.; BRISKEY, E. J.; CASSENS, R. G.; FORREST, J. C.; MEYER, R. K. (1968): Adrenal and thyroid function in stress-susceptible pigs (*Sus domesticus*). Amer. J. Physiol. 214, 146-151.
KALBITZ, S. (1986): Untersuchungen der Insulinrezeptoren und des Glukosemetabolismus im Rückenspeck von Schweinen verschiedener Zuchtlinien. Dissertation, Universität Hohenheim, Stuttgart, GFR.
KATTESH, H. G.; KORNEGAY, E. T.; KNIGHT, J. W.; GWAZDAUSKAS, F. G.; THOMAS, H. R.; NOTTER, D. R. (1980): Glucocorticoid concentrations, corticosteroid binding protein characteristics and reproduction performance of sows and gilts subjected to applied stress during mid-gestation. J. Anim. Sci. 50, 897-905.
KELLER-WOOD, M. E.; DALLMAN, M. F. (1984): Corticosteroid inhibition of ACTH secretion. Endocrinol. Rev. 5, 1-24.
KING, B. M.; SMITH, R. L. (1985): Hypothalamic obesity after hypophysectomy or adrenalectomy: dependence on corticosterone. Amer. J. Physiol. 249, 522-526.
LEVINE, S. (1985): A definition of stress? In: Animal Stress. (Editor: MOBERG, G.P.). Amer. Physiol. Soc., Bethesda, Maryland, USA. Pp. 51-69.
LISTER, D.; HALL, G. M.; LUCKE, J. N. (1976): Porcine malignant hyperthermia. 3. adrenergic blockade. Brit. J. Anaesth. 48, 831-837.
LUCKE, J. N.; DENNY, H.; HALL, G. M.; LOVELL, R.; LISTER, D. (1978): Porcine malignant hyperthermia. 6. The effects of bilateral adrenalectomy and pretreatment with bretylium on the halothane-induced response. Brit. J. Anaesth. 50, 241-245.
LUCKE, J. N.; HALL, G. M.; LISTER, D. (1976): Porcine malignant hyperthermia. 1. Metabolic and physiological changes. Brit. J. Anaesth. 48, 297-304.
LUCUS, C. P.; ESTIGARRIBIA, J. A.; DARGA, L. L.; REAVEN, G. M. (1985): Insulin and blood pressure in obesity. Hypertension 7, 702-706.
MABRY, J. W.; CHRISTIAN, L. L.; KUHLERS, D. L. (1981): Inheritance of porcine stress syndrome. J. Heredity 72, 429-430.
MABRY, J. W.; CHRISTIAN, L. L.; KUHLERS, D. L.; RASMUSEN, B. A. (1983): Prediction of susceptibility to the porcine stress syndrome. J. Heredity 74, 23-26.
MAINS, R. E.; EIPPER, B. A.; LING, N. (1977): Common precursor to corticotropins and endorphins. Proc. Natl. Acad. Sci. USA 74, 3014-3018.
MARKS, N. (1978): Biotransformation and degradation of corticotropins, lipotropins and hypothalamic peptides. In: Frontiers in Neuroendocrinology Vol. 5. (Editors: GANONG, W. F.; MARTINI, L.). Raven Press, New York. Pp. 329-353.
MARPLE, D. N.; ABERLE, E. D.; FORREST, J. C.; BLAKE, W. H.; JUDGE, M. D. (1972b): Endocrine response of stress susceptible and stress resistant swine to environmental stressors. J. Anim. Sci. 35, 576-579.
MARPLE, D. N.; CASSENS, R. G. (1973a): Increased metabolic clearance of cortisol by stress-susceptible swine. J. Anim. Sci. 36, 1139-1142.
MARPLE, D. N.; CASSENS, R. G. (1973b): A mechanism for stress-susceptibility in swine. J. Anim. Sci. 37, 546-550.
MARPLE, D. N.; CASSENS, R. G.; TOPEL, D. G.; CHRISTIAN, L. L. (1974a): Porcine corticosteroid-binding globulin: binding properties and levels in stress-susceptible swine. J. Anim. Sci. 38, 1224-1228.
MARPLE, D. N.; JUDGE, M. D.; ABERLE, E. D. (1972a): Pituitary and adrenocortical function of stress susceptible swine. J. Anim. Sci. 35, 995-1006.
MARPLE, D. N.; TOPEL, D. G.; MATSUSHIMA, C. Y. (1969): Influence of induced adrenal insufficiency and stress on porcine plasma and muscle characteristics. J. Anim. Sci. 29, 882-886.
MÜLLER, E.; ROGDAKIS, E. (1985): Genetische Regulation des Fettstoffwechsels beim Schwein. In: Methodische Ansätze in der Tierzüchtung. In: Hohenheimer Arbeiten 131, Verlag Ulmer, München, GFR. Pp. 8-28.
MUNCK, A.; GUYRE, M.; HOLBROOK, N. J. (1984): Physiological functions of glucocorticoids in stress and their relation to pharmacological actions. Endocrinol. Rev. 5, 25-44.
NATIONAL DIABETES DATA GROUP (1979): Classification and diagnosis of diabetes and other categories of glucose intolerance. Diabetes 28, 1039-1057.
OHNISHI, S. T. (1987): Effects of halothane, caffeine, dantrolene and tetracaine on the calcium permeability of skeletal sarcoplasmic reticulum of malignant hyperthermic pigs. Biochim. Biophys. Acta 897, 261-268.
OHNISHI, S. T.; WARING, A. S.; FANG, S. R. G.; HORIUCHI, K.; FLICK, S. L.; SODANAGA, K. K.; OHNISHI, T. (1986): Abnormal membrane properties of the sarcoplasmic reticulum of pigs susceptible to malignant hyperthermia: modes of action of halothane, caffeine, dantrolene, and two other drugs. Arch. Biochem. Biophys. 247, 294-301.
O'RALILLY, S.; WAINSCOAT, J. S.; TURNER, R. C. (1988): Type 2 (non-insulin-dependent) diabetes mellitus. New genetics for old nightmares. Diabetologia 31, 407-414.

Patty, M.; Schlesinger, D. H.; Horvath, J.; Mason-Garcia, M.; Szoke, B.; Schally, A. V. (1986): Purification and characterization of peptides with corticotropin- releasing factor activity from porcine hypothalami. Proc. Natl. Acad. Sci. USA **83**, 2969-2973.

Reaven, G. M.; Chen, Y. D. I.; Golay, A.; Swislocki, A. L. M.; Jaspan, J. B. (1987): Documentation of hyperglucagonemia throughout the day in non-obese and obese patients with noninsulin-dependent diabetes mellitus. J. Clin. Endocrinol. Metab. **64**, 106-110.

Reaven, G. M.; Hoffman, B. B. (1987): A role for insulin in the aetiology and course of hypertension. Lancet **2**, 435-436.

Reaven, G. M.; Hoffmann, B. B. (1988): Abnormalities of carbohydrate metabolism may play a role in the etiology and clinical course of hypertension. Trends Pharmacol. Sci. **9**, 78- 79.

Reik, T. R.; Rempel, W. E.; McGrath, C. J.; Addis, B. (1983): Further evidence on the inheritance of halothane reaction in pigs. J. Anim. Sci. **57**, 826-831.

Richter, W. O.; Naude, R. J.; Oelofson, W.; Schwandt, P. (1987): In vitro lipolytic activity of beta-endorphin and its partial sequences. Endocrinology **120**, 1472- 1476.

Richter, W. O.; Schwandt, P. (1987): Lipolytic potency of proopiomelanocorticotropin peptides in vitro. Neuropeptides **9**, 59-74.

Rivier, C.; Brownstein, M.; Spiess, J.; Rivier, J.; Vale, W. (1982a): In vivo corticotropin-releasing factor-induced secretion of adrenocorticotropin, beta- endorphin, and corticosterone. Endocrinology **110**, 272-278.

Rivier, C.; Rivier, J.; Mormède, P.; Vale, W. (1984a): Studies of the nature of the interaction between vasopressin and corticotropin-releasing factor on adrenocorticotropin release in the rat. Endocrinology **115**, 882-886.

Rivier, C.; Rivier, J.; Vale, W. (1982b): Inhibition of adrenocorticotropic hormone secretion in the rat by immunoneutralization of corticotropin-releasing factor. Science **218**, 377-379.

Rivier, J.; Rivier, C.; Vale, W. (1984b): Synthetic competitive antagonists of corticotropin-releasing factor: effect on ACTH secretion in the rat. Science **224**, 889-891.

Rivier, C.; Vale, W. (1983): Interaction of corticotropin-releasing factor and arginine vasopressin on adrenocorticotropin secretion in vivo. Endocrinology **113**, 939-942.

Rogdakis, E. (1982): Selektion nach der Aktivität NADPH-liefernder Enzyme im Fettgewebe des Schweines. 1. Versuchsfrage, Versuchsanlage und erste Ergebnisse. Z. Tierz. Züchtungsbiol. **99**, 241-252.

Rogdakis, E.; Ensinger, U.; Faber, H.,v. (1979): Hormonspiegel im Plasma und Enzymaktivitäten im Fett-gewebe von Piétrain- und Edelschweinen. Z. Tierz. Züchtungsbiol. **96**, 108-119.

Rogdakis, E.; Haid, H.; Faber, H.,v. (1975): Endogene 11-Hydroxykortikosteroide beim Piétrain- und Edelschwein sowie ihren Kreuzungsprodukten und ihre Beziehungen zur Fleischqualität. Züchtungskunde **47**, 311-318.

Salati, L. M.; Adkins, B.; Clarke, S. D. (1988): Free fatty acid inhibition of the insulin induction of glucose-6-phosphate dehydrogenase in rat hepatocyte monolayers. Lipids **23**, 36-41.

Schmale, H.; Richter, D. (1984): Single base deletion in the vasopressin gene is the cause of diabetes insipidus in Brattleboro rats. Nature **308**, 705-709.

Schneebeli, H. (1987): Metaboliten und Hormone bei Schweinen unterschiedlicher Leistung und Körperzusammensetzung. Thesis No. 8179, ETH Zürich, Switzerland.

Selye, H. (1976): Stress in Health and Disease. Butterworths, Boston, USA.

Strutz, C.; Rogdakis, E. (1979): Phenotypic and genetic parameters of NADPH- generating enzymes in porcine adipose tissue. Z. Tierz. Züchtungsbiol. **96**, 170- 185.

Takahashi, E. (1986): Long-term blood-sampling technique in piglets. Lab. Anim. **20**, 206-209.

Vale, W.; Spiess, J.; Rivier, C.; Rivier, J. (1981): Characterization of a 41-residue ovine hypothalamic peptide that stimulates secretion of corticotropin and beta-endorphin. Science **213**, 1394-1397.

Vogeli, P.; Gerwig, C. (1979): Indexselektion beim Schwein auf zwei entgegengesetzte Zuchtlinien. 1. Mitteilung: Erzielter und erwarteter Zuchtfortschritt bei den wesentlichsten Leistungsmerkmalen. Schweiz. Landw. Monatshefte **57**, 257-279.

Vogeli, P.; Gerwig, C.; Schneebeli, H.; Wäfler, P. (1982): Genfrequenzen wichtiger polymorpher Systeme des Veredelten Landschweines bei Indexselektion in positiver und negativer Richtung. Schweiz. Landw. Monatshefte **60**, 234-240.

Vogeli, P.; Stranzinger, G.; Schneebeli, H.; Hagger, C.; Künzi, N.; Gerwig, C. (1984): Relationship between the H and A-O blood types, phosphohexose isomerase and 6- phosphogluconate dehydrogenase red cell enzyme systems and halothane sensitivity, and economic traits in a superior and an inferior selection line of Swiss landrace pigs. J. Anim. Sci. **59**, 1440-1449.

Wäfler, P.; Schneebeli, H.; Gerwig, C.; Blum, J. (1982): Wirkungen eines Indexes bezüglich erwünschter und unerwünschter Selektionseffekte beim Schwein. Schweiz. Landw. Monatshefte **60**, 241-247.

Webb, A. J. (1981): The halothane sensitivity test. In: Porcine Stress and Meat Quality - Cause and Possible Solutions to the Problem. Proc. Symp. Refsnes. Gods, Norway. Pp. 105-124.

ZADINA, J. E.; BANKS, W. A.; KASTIN, A. J. (1986): Central nervous system effects of peptides, 1980-1985. A cross-listing of peptides and their central actions from the first six years of the journal Peptides. Peptides **7**, 497-537.

Authors' addresses:
J.J. Hari, Merck Sharp & Dohme-Chibret AG, Schaffhauserstr. 136, CH-8152 Glattbrugg, Switzerland;
E. Lang, Institut für Nutztierwissenschaften, ETH Zürich, CH-8092 Zürich, Switzerland.

Novel variants of the stress hormone ACTH in pigs: genetic and pathophysiological considerations.

K. VOIGT, W. STEGMAIER and G. MCGREGOR

Background

Hormonal and neural peptides are a large and varied group of secretory proteins which have a regulatory function in diverse physiological processes. They are all generated from larger precursor proteins by endoproteolysis sometimes followed by further post-translational processing (ANDREWS et al. 1987, WOLD 1981). Much of our present knowledge of the features of peptide biosynthesis has come from studies employing molecular biological methods. However, information about the structure of the final product, the biologically active peptide, requires its isolation, purification and direct biochemical characterization. We have taken this approach to study the pattern of biosynthetic processing of corticotropin (ACTH) in the pituitary gland of the pig. ACTH is a 39 amino acid peptide derived from the precursor polypeptide, proopiomelanocortin (POMC) by endoproteolytic cleavage at double basic residues which flank the peptide sequence within the precursor (MAINS et al. 1977, ROBERTS & HERBERT 1977, CHRÉTIEN et al. 1979). We were able to identify 6 variant forms of ACTH. Four of these may be derived from ACTH by post-translational processing. The other two variant forms represent a novel primary structure for porcine ACTH. The physiological and genetic implications of this will be discussed.

The major physiological function of ACTH appears to be in the response to stress by stimulating the secretion of cortisol from the adrenal cortex. However, it appears to possess additional intrinsic biological activity. SCHWYZER (1980) has delineated domains within the primary structure of ACTH (1-24) to which specific functional properties can be assigned. Such a detailed characterization of structure-function relationships has so far not been possible for any other hormonal or neural peptide. Differential processing of ACTH could provide different biological signals. For example, it is already known that in the neurointermediate lobe of the pituitary intact ACTH(1-39) is processed further to α-MSH (N-α-acetyl-ACTH(1-13) amide) and CLIP (corticotropin-like intermediate lobe peptide, corresponding to ACTH(18-39)) (SCOTT et al. 1973, MAINS & EIPPER 1981, ROBERTS et al. 1978). Neither of these peptides possess the steroidogenic activity of ACTH (1-39) but possess distinctly different biological properties. In contrast, in the anterior pituitary ACTH(1-39) is formed. Also ACTH(1-38) has been identified in pituitary extracts (BRUBAKER et al. 1980). In addition, ACTH(7-39) has been isolated from pituitary gland of pig (EKMAN et al. 1984), human (LI et al. 1978) and elephant (LI et al. 1988). Interestingly, this peptide has been shown to act antagonistically to ACTH(1-39) (LI et al. 1978). This differential processing of ACTH offers a mode of regulating endocrine function. However, the precise regulatory mechanisms are not yet clear but tissue-specific factors may be important.

The sequence of POMC from six mammalian species: ox (NAKANISHI et al. 1979), pig (BOILEAU et al. 1983a, GOSSARD et al. 1986), rat (DROUIN et al. 1985), mouse (NOTAKE et al. 1983), monkey (PATEL et al. 1988) and human (TAKAHASHI et al. 1981), and two non-mammalian species: *Xenopus* (MARTENS 1986) and salmon (KAWAUCHI 1983) have now been determined by sequencing cloned cDNA derived from mRNA extracted from the pituitary gland of these species (Fig. 1). There is a remarkable sequence conservation especially within the amino-terminal region of ACTH. The sequence of ACTH(2-12) is identical across all these species including the two non-mammalian species. The sequence of ACTH(1-24) is identical in all 6 mammalian species and has only minor differences at 4 positions in the *Xenopus* and salmon sequences. This would suggest that ACTH(1-24) has an important biological function. Furthermore, it seems evident that the complete fidelity of the primary structure of ACTH(1-24) is a strict biological requirement. This suggests that even a single residue change within the sequence is not tolerated. This is consistent with the existence of a precise functional organization as has been suggested for this sequence.

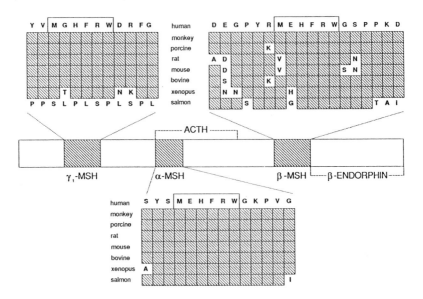

Fig. 1. The primary structure of POMC has been determined for 6 mammalian and 2 non-mammalian species by sequencing cloned cDNA. In all species, the sequence Met-Glu/Gly-His-Phe-Arg-Trp is repeated and is within the hatched areas which correspond to α, β, and γ_1-MSH. These regions of POMC have been highly conserved through evolution. One letter symbols of amino acids according to the IUPAC-IUB Joint Commission on Biochemical Nomenclature, cf. Eur. J. Biochem. **138** (1984), 9-37.

Isolation of ACTH variants

We extracted pig pituitary glands according to the method of SCHLEYER et al. (1969) by which four fractions with lipolytic activity are identified. One of these fractions was known to contain a considerable amount of ACTH activity and therefore, was taken as our starting material. Initial chromatography on Sephadex G75 resolved two major peaks of UV absorbance. This was in accordance with SCHLEYER and coworkers (1969) who found that most of the ACTH activity was recovered in the second peak. This fraction was further resolved into three peaks on gel permeation chromatography on Fractogel TSK HW-40. Most of the ACTH activity was recovered in the third of these peaks which was then fractionated on reverse-phase (r.p.) HPLC. The major UV absorbing fractions obtained from the first r.p. HPLC step were each then separately purified by sequential r.p. HPLC steps. This was achieved by changing the selectivity of consecutive chromatographic steps by switching either the type of reverse-phase matrix or the counter-ion of the mobile phase (Fig. 2). The purified fractions were then analysed to obtain their primary structures. For this purpose we combined extensive peptide mapping with micromethods of amino acid analysis using OPA derivatization (CHEN et al. 1979) and manual sequence determination by the DABIT/PITC-Method (see WITTMANN-LIEBOLD et al. 1984). The primary sequences of the different fractions are shown in Fig. 3.

Our finding of ACTH(1-38) in pig pituitary extracts confirms the finding of EKMAN et al. (1984) who in addition isolated ACTH(1-37) and (7-39). We failed to identify ACTH(1-37) and ACTH(7-39) and found five novel variants of ACTH. EKMAN's group and ourselves started with different prepurified subfractions of porcine pituitary extracts and this may account for the differences in variant forms isolated. Conversely, these differences may reflect differences in the pattern of processing between different strains of pig. The understanding of the biological significance of the

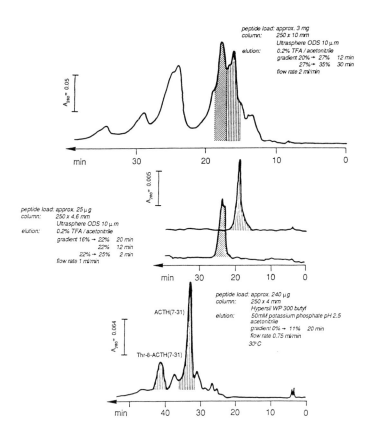

Fig. 2. This example illustrates the general scheme employed for purification of the ACTH variants. Reverse-phase HPLC columns were employed and the eluates monitored for UV absorbance in order to identify the major peptide-containing fractions. The fractions corresponding to each of the shaded regions of the upper profile were pooled. These were then separately re-chromatographed under different reverse-phase HPLC conditions as indicated. The profiles of UV absorbance so obtained are the middle profiles shown here. Each of the shaded peaks were re-run and the profile obtained for one of these (the vertically hatched peak) is shown in the bottom profile. Here it is seen that a further change in chromatographic conditions allowed the resolution of two major peaks. These (vertically hatched) peaks were taken for sequence analysis which revealed their identity as the forms of ACTH which are indicated.

existence of these truncated ACTH variants requires full biological testing of synthetic versions of these fractions. These sequences are closely related to ACTH and appear to represent truncated versions of ACTH(1-39). But, in addition, two of these peptides contain a variant of the primary structure, containing threonine (instead of arginine) at position 8. This represents the first reported variant of the ACTH primary structure.

Biological activity of ACTH variants

We have assayed the biological activity of four of our ACTH variant preparations with respect to their corticotrophic and lipolytic activities compared to that of ACTH(1-39). The corticotrophic

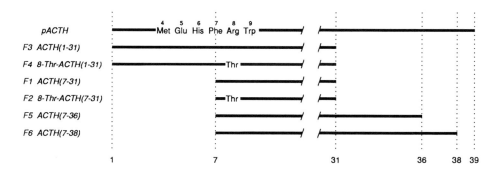

Fig. 3. Diagram of essential structural features of the variants of ACTH in comparison to porcine ACTH(1-39). Each of them represents a truncated form of ACTH and the carboxy- and amino-terminal residues are indicated with the number of the corresponding amino acid residue in ACTH(1-39). In addition, fractions F2 and F4 contain a threonine, as shown in the position corresponding to position 8, and this is the only different primary structure found in these variants compared to that of ACTH(1-39).

activity was assessed by measuring corticosterone release from isolated rat adrenocortical cells (SAYERS et al. 1971), and lipolytic activity assayed by measuring glycerol release from isolated fat cells (SCHLEYER et al. 1969). The comparative steroidogenic potency of four of our variants and porcine ACTH(1-39) is illustrated in Fig. 4. All exhibited less activity in these systems than ACTH(1-39). With respect to corticotrophic and lipolytic activities the absence of the amino terminus of ACTH has a profound effect on biological potency. All such variants had 1% or less of the activity of ACTH(1-39). This contrasts with 30%-40% for ACTH(1-31). However, substitution by threonine (Thr) in position 8 has a more profound effect on biological activity: Thr-8-ACTH(1-31) is a further order of magnitude less active than ACTH(1-31) and Thr-8-ACTH(7-31) has less than 1% of the

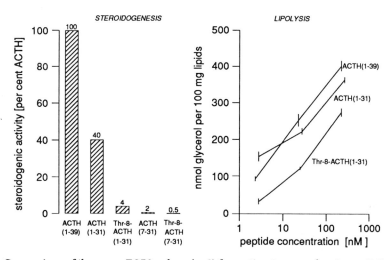

Fig. 4. Comparison of the mean EC50 values (n=6) for corticosterone-releasing activity of four of the ACTH variants compared to porcine ACTH(1-39) is shown in the histogram (left hand panel). The relative values are expressed as a percentage of the steroidogenic activity of porcine ACTH(1-39). Similar relative potencies were found for lipolytic activity and the results obtained for three of the peptides are shown in the right hand panel.

activity of ACTH(1-39). Coincubation of the variants Thr-8-ACTH(1-31) or ACTH(7-31) with natural ACTH(1-39) did not disclose any antagonistic effect on corticotrophic activity.

Possible biosynthetic origins of ACTH variants

The origin of these variants needs to be explained. If they were all derived from the ACTH(1-39) sequence during the biosynthetic processing of POMC, novel proteolytic processing is involved. The possibility that each sequence is derived not from ACTH but from another precursor must also be considered.

The isolation of the 8-threonine analogues, Thr-8-ACTH(7-38) and Thr-8-ACTH(1-31), is especially interesting. They both may be derived from the same precursor which is distinct from POMC and which contains this ACTH-like sequence. Such homology could originate from gene duplication and subsequent structural divergence including the change at the position corresponding to 8-arginine of ACTH. However, there is no direct evidence for a second porcine gene. In both *Xenopus* (MARTENS 1986) and salmon (KAWAUCHI 1983) two forms of POMC have been identified which differ in primary structure and so indicate the existence of a second POMC gene. It is possible that in mammals a duplicate POMC gene is present but has escaped detection having undergone significant divergence whilst retaining the ACTH-like sequence. The concomitant acquisition of different biological properties would also be expected. Alternatively, the ACTH variants are derived from a mutated form of the POMC gene present in a subpopulation of the pigs used here. It is also possible that such a mutation has occurred in only one of the alleles of the POMC gene. This would parallel findings of insulin variants in the human population (SHOELSON et al. 1983). It is interesting to note that mRNA purified from pig pituitary glands generates by *in vitro* translation two forms of POMC (BOILEAU et al. 1983b). These may arise from separate non-allelic genes but, as BOILEAU discusses, "... the possibility of two allelic variants in the hog population studied exists".

Possible pathophysiologigal relevance of ACTH variants

Single amino acid substitutions have been identified in human insulin and proinsulin. The clinical consequences of these mutations have been described and serve to illustrate the potential significance to health of such changes. STEINER and coworkers (SHOELSON et al. 1983) have reported three human diabetic individuals each possessing a different mutant form of insulin. The mutations were at different sites but each resulted in a single amino acid substitution. Also, they all presented similar symptoms - hyperglycemia and hyperinsulinemia. In each case it was possible to locate the mutant to just one of the insulin alleles. The tissue ratio of normal to mutant insulin was close to 1:1. However, less than 5% of the circulating insulin was of the normal form. It was concluded that the mutant insulin, which had reduced biological activity, was inefficiently cleared from the circulation. This leads to reduced levels of normal insulin and consequently, to hyperglycemia. Families and individuals have been identified with elevated levels of proinsulin (ROBBINS et al. 1981, SHIBASAKI et al. 1985). This has been found to be due to a mutation at a single arginine residue - the site of proteolytic cleavage of the proinsulin by which insulin is generated. The mutation blocks this conversion. Similar mutations, also leading to an obstruction of proteolytic activation, have been identified in proalbumin; elevation of circulating proalbumin levels has been observed in this instance (BRENNAN & CARELL 1978).

Therefore, a single amino acid substitution within a hormone or neuropeptide may have profound effects on endocrine, neuroendocrine or neural functions. This may occur without the mutation directly altering the domain responsible for intrinsic activity. The significant consequences of a single amino acid substitution may be due to altered biosynthetic or catabolic processing.

The finding of variant forms of human insulin suggests that similar mutations may arise in other hormonal and neuropeptide genes with similar frequency. This has important implications for human and animal disease. The true extent of such abnormalities will only be revealed by further extensive investigations in which the methods of peptide isolation and characterization will be of key importance.

A mutational change at position 8 of ACTH might therefore be expected to have profound biological consequences. It is within the sequence ACTH(5-9) (normally Glu-His-Phe-Arg-Trp),the one assigned by SCHWYZER (1980) as the core of the peptide domain which directly interacts with the receptor on steroid-producing cells of the adrenal cortex. In addition, this pentapeptide sequence occurs at two other sites within each of the known mammalian POMCs and also in corresponding positions (Fig. 4). This is also true for *Xenopus* POMC and in salmon the sequence occurs at one of these additional sites. Such replication of biologically active peptide sequences within a single precursor polypeptide is very common and is considered to be a measure of the biological importance of the structure.

It is clear from our bioassay data that a substitution at position 8 of the ACTH sequence crucially effects corticotrophic and lipolytic activity. It might be expected that individual pigs expressing such a mutant form of ACTH would exhibit an abnormal physiological response to stress. "Porcine Stress Syndrome" is a well recognized condition found in subpopulations of several strains of domestic pig (see GORDON et al. 1973). Pigs with this condition elicit an exaggerated stress response to certain environmental stimuli. The primary disorder is not yet identified, but evidently, it has a genetic basis. The possibility of a connection between this disorder and the expression of a mutant form of ACTH seems to be worth investigating.

Concluding remarks

Further studies are required to establish the genetic basis of our finding of a novel ACTH sequence. The biosynthesis of ACTH and other products of POMC have been extensively studied by several groups and the pig has often been the species of choice. Therefore, it is perhaps surprising that our "mutant" sequence has not been previously identified and confirmed by other workers. It may be that given the rapid and artificial selection of domestic animals, this mutant was only a transitory occurrence in the pig population. However, the possibility that the finding has a relevance for the health of pigs deserves consideration, particularly since human diseases caused by single amino acid substitutions in endocrine peptides have become apparent.

Acknowledgement

The research was supported by Deutsche Forschungsgemeinschaft (DFG, grant Vo 253/6-1).

References

ANDREWS, P.C.; BRAYTON, K.; DIXON, J.E. (1987): Precursor to regulatory peptides: their proteolytic processing. Experientia **43**, 784-790.

BOILEAU, G.; BARBEAU, C.; JEANNOTTE, L.; CHRÉTIEN, M.; DROUIN, J. (1983a): Complete structure of the porcine pro-opiomelanocortin mRNA derived from the nucleotide sequence of cloned cDNA. Nucl. Acids Res. **11**, 8063-8071.

BOILEAU, G.; GOSSARD, F.; SEIDAH, N.G.; CHRÉTIEN, M. (1983b): Cell-free synthesis of porcine proopiomelanocortin: two distinct primary translation products. Can. J. Biochem. Cell Biol. **61**, 333-339.

BRENNAN, S.O.; CARELL, R.W. (1978): Circulatory variant of human proalbumin. Nature **274**, 908-909.

BRUBAKER, P.L.; BENNETT, H.P.J.; BAIRD, A.C.; SOLOMON, S. (1980): Isolation of ACTH(1-39), ACTH(1-38) and CLIP from the calf anterior pituitary. Biochem. Biophys. Res. Commun. **96**, 1441-1448.

CHEN, R.F.; SCOTT, C.; TREPMAN, E. (1979): Fluorescence properties of o-phthaldialdehyde derivatives of amino acids. Biochim. Biophys. Acta **576**, 440-455.

CHRÉTIEN, M.; BENJANNET, S.; GOSSARD, F.; GIANOULAKIS, C.; CRINE, P.; LIS, M.; SEIDAH, N.G. (1979): From beta-lipotropin to beta-endorphin and 'pro-opio-melanocortin'. Can. J. Biochem. **57**, 1111-1121.

DROUIN, J.; CHAMBERLAND, M.; CHARRON, J.; JEANNOTTE, L.; NEMER, M. (1985). Structure of the rat pro-opiomelanocortin (POMC) gene. FEBS Lett. **193**, 54-58.

EKMAN, R.; NORÉN, H.; HAKANSON, R.; TÖRNVALL, H. (1984). Novel variants of adrenocorticotrophic hormone in porcine anterior pituitary. Regul. Peptides **8**, 305-314.

GORDON, R.N.; BRIT, B.A.; KALOW, W. (1973): 1st International Symposium on Malignant Hyperthermia, Toronto, 1971. Charles C. Thomas, Springfield, Ill., USA.

Gossard, F.J.; Chang, A.C.Y.; Cohen, S.N. (1986): Sequence of the cDNA encoding porcine pro-opiomelanocortin. Biochim. Biophys. Acta **866**, 68-74.

Kawauchi, M. (1983): Chemistry of proopiomelanocortin-related peptides in the salmon pituitary. Arch. Biochem. Biophys. **227**, 343-350.

Li, C.H.; Chung, D.; Yamashiro, D.; Lee, C.Y. (1978): Isolation, characterization, and synthesis of a corticotropin-inhibiting peptide from human pituitary glands. Proc. Natl. Acad. Sci. USA **75**, 4306-4309.

Li, C.H.; Oosthuizen, M.M.J.; Chung, D. (1988): Isolation and primary structures of elephant adrenocorticotropin and beta-lipotropin. Int. J. Peptide Protein Res. **32**, 573-578.

Mains, R.E.; Eipper, B.A. (1981): Differences in the post-translational processing of ß-endorphin in rat anterior and intermediate pituitary. J. Biol. Chem. **256**, 5683-5688.

Mains, R.E.; Eipper, B.A.; Ling, N. (1977): Common precursor to corticotrophins and endorphins. Proc. Natl. Acad. Sci. USA **74**, 3014-3018.

Martens, G.J.M. (1986): Expression of two proopiomelanocortin genes in the pituitary gland of *Xenopus laevis*: complete structures of the two preprohormones. Nucl. Acids Res. **14**, 3791-3798.

Nakanishi, S.; Inoue, A.; Kita, T.; Nakamura, M.; Chang, A.C.Y.; Cohen, S.N.; Numa, S. (1979): Nucleotide sequence of cloned cDNA for bovine corticotropin-ß-lipotropin precursor. Nature **278**, 423-427.

Notake, M.; Tobimatsu, I.; Watanabe, I.; Takahashi, H.; Mishina, M.; Numa, S. (1983): Isolation and characterization of the mouse corticotropin-ß-lipotropin precursor gene and a related pseudogene. FEBS Lett. **156**, 67-71.

Patel, P.D.; Sherman, T.G.; Watson, S.J. (1988): Characterization of pro-opiomelanocortin cDNA from Old World monkey, *Macaca nemestrina*. DNA **7**, 627-635.

Robbins, D.C.; Blix, P.M.; Rubenstein, A.H.; Kanazawa, Y.; Kosaka, K.; Tager, H.S. (1981): Human proinsulin variant at arginine 65. Nature **291**, 679-681.

Roberts, J.L.; Herbert, E. (1977): Characterization of a common precursor to corticotropin and ß-lipotropin: Cell-free translation of the precursor and identification of corticotropin peptides in the molecule. Proc. Natl. Acad. Sci. USA **74**, 4826-4830.

Roberts, J.L.; Phillips, M.; Rosa, P.A.; Herbert, E. (1978): Steps involved in the processing of common precursor forms of adrenocorticotropin and endorphin in cultures of mouse pituitary cells. Biochemistry **17**, 3609-3618.

Sayers, G.; Swallow, R.L.; Giordano, N.D. (1971): An improved technique for the preparation of isolated rat adrenal cells: a sensitive and specific method for the assay of ACTH. Endocrinology **88**, 1063-1068.

Schleyer, M.; Straub, K.; Faulhaber, J.D.; Pfeiffer, E.F. (1969): Studies on the pituitary "Fettstoffwechselhormon". II. The lipolytic activity of various pituitary gland fractions *in vitro*. Horm. Metab. Res. **1**, 286-289.

Schwyzer, R. (1980): Organization and transduction of peptide information. Trends Pharmacol. Sci. **1**, 327-331.

Scott, A.P.; Ratcliffe, J.G.; Rees, L.H.; Landon, J.; Bennett, H.P.J.; Lowry, P.J.; McMartin, C. (1973): Pituitary Peptide. Nature New Biol. **244**, 65-67.

Shibasaki, Y.; Kawakami, T.; Kanazawa, Y.; Akanuma, Y.; Takaku, F. (1985): Posttranslational cleavage of proinsulin is blocked by a point mutation in familial hyperproinsulinemia. J. Clin. Invest. **76**, 378-380.

Shoelson, S.; Haneda, M.; Blix, P.; Nanjo, A.; Sanke, T.; Inouye, K.; Steiner, D.; Rubenstein, A.; Tager, H. (1983): Three mutant insulins in man. Nature **302**, 540-543.

Takahashi, H.; Teranishi, Y.; Nakanidhi, S.; Numa, S. (1981): Isolation and structural organization of the human corticotropin-ß-lipotropin precursor gene. FEBS Lett. **135**, 97-102.

Whitfield, P.L.; Seeburg, P.H.; Shine, J. (1982): The human pro-opiomelanocortin gene: Organization, sequence, and interspersion with repetitive DNA. DNA **1**, 133-143.

Wittmann-Liebold, B.; Kimura, M.; Walker, J.M. (Editors) (1984): Methods in Molecular Biology. Humana Press, Clifton.

Wold, F. (1981): *In vivo* chemical modification of proteins (post-translational modification). Ann. Rev. Biochem. **50**, 783-814.

Authors' address:
K.-H. Voigt, W. Stegmaier and G. McGregor, Universität Marburg, Institut für Physiologie, Deutschhausstrasse 1-2, D-3550 Marburg, F.R.G.

Catecholamine determination in pig peripheral plasma as an alternative parameter to characterize stress situations

M. DEHNHARD and R. CLAUS

Stress syndrome

Each organism must be able to react to challenges of the environment in order to survive. Such deviations from normal environmental conditions are termed "stressors" and include high or low temperatures, thirst, infection, injury, excessive muscular activity and abnormal psychical situations such as fear. The organism shows a variety of physiological and behavioral reactions to counteract stressors. Depending on the duration of the stressful situation, the biological responses are usually divided into two syndromes. The "fight and flight syndrome" (CANNON & DE LA PAZ 1911) is the immediate reaction in response to a sudden and short-term critical situation. The consequences may be, for instance, an increase of the heart frequency and the blood pressure, an activation of the glucose supply by an enhanced degradation of glycogen by the liver (glycogenolysis), and an additional energy supply by lipolysis. The second is the "general adaptation syndrome" (SELYE 1936) which is the long-term response to continuous stressors. It may include morphological changes, such as adrenal or lymphatic hypertrophy.

Animal husbandry has to provide environmental and management conditions which fit to the demands of a high productivity and an optimal animal welfare as well. This leads to the search for parameters which objectively seperate favourable from unfavourable conditions and moreover allow to quantify the extent of an unfavourable situation. It is obvious that the physiological reactions which are involved in the stress syndromes are possible candidates for parameters which allow quantification of stress.

Quantification of stress

Cardiovascular and metabolic parameters

Changes of heart rate and blood pressure are obvious within seconds after exposure to physical exercise. In boars, physical "stress" such as mating, leads to a more than 2.5-fold increase in the heart rate (77 vs. 196 min^{-1}) and a 1.6-fold increase in both the systolic and diastolic blood pressure (PEREZ et al. 1988).

Metabolic parameters may include enzyme activities involved in the supply of energy such as creatine kinase and metabolites such as lactate. Direct conclusions on the energy supply from the adipose tissue can be obtained from the measurement of free fatty acids (FFA) in blood. It is well established that in case of fasting, the concentrations of FFA are markedly increased. Apparently a dose dependency also exists, since a moderate reduction in energy supply (30%) is reflected by moderately increased FFA (Fig.1). Nervous control of adipose tissue and the hormones involved in lipolysis suggest that FFA are susceptible to physical as well and to emotional stress. In sheep it was shown, that emotional and physical consequenes of experimental procedures lead to a 1.5-fold elevation of FFA concentrations (SLEE & HALLIDAY 1968). This increase, however, is much higher after acute cold exposure (-18° C) which led to a more than 10-fold elevation of FFA (2 mequiv/l vs. 0.1–0.2 mequiv/l) to provide the increased energy requirement. In this study large differences between individuals in the maximum concentrations were observed (range 0.68 to 4.23 mequiv/l, n=100). Similarly, in cows exposed to -19°C to -26°C, the FFA concentrations were significantly increased (OLSON & TRENKLE 1973).

Emotional stress and exercise are also potent stimuli to increase FFA concentrations in cows and pigs. In cows the effect of transportation on FFA levels was demonstrated by REYNAERT et al.(1976). Both loading and transportation led to a more than threefold increase in serum FFA. In halothane

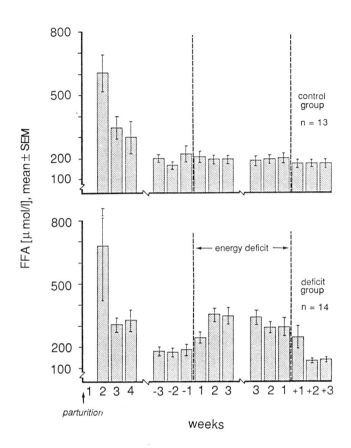

Fig.1. Concentration of FFA (µmol COOH / l blood plasma) in dairy cows post partum (arithmetic mean ± SEM). Controls were fed according to maintenance plus milk production requirements from parturition until the end of the investigation (upper panel). In contrast, in the animals of the "deficit group" an energy deficit (30%) was provoked for a period of about 2 consecutive estrous cycles (lower panel) (SCHOPPER & CLAUS, in preparation).

positive boars a standardized stress (running belts) was applied, until the heart rate was raised up to 270 min^{-1}. The resulting concentrations of FFA in plasma were more than 2-fold (0.183 vs. 0.428 mequiv/l). In halothane negative boars subjected to the same treatment the corresponding FFA concentrations were 0.263 and 0.575 mequiv/l respectively (LENGERKEN et al. 1979). The differences between both groups, however, were not significant. These results do not support the hypothesis that the halothane reaction is a parameter for general stress susceptibility.

Glucocorticoids

The alterations of heart rate, blood pressure and the mobilization of FFA described above occur as a consequence of a number of neurohormonal responses during stress. Two principle systems are involved in this response: the sympathetic-adrenomedullary system (release of noradrenaline and adrenaline by the adrenal medulla) and the hypothalamic-pituitary-adrenocortical system (e.g. secretion of glucocorticoids from the adrenal cortex). It is known that interactions occur between the two systems (OLIVERIO 1987). For corticosteroids very close time relationships between stress exposure and the concentrations in plasma were shown. They increase within minutes after the onset of different types of stress. Physical and emotional stress, such as loading pigs onto a lorry resulted in an 2-fold elevation of plasma corticosteroids (4.3 vs. 9.1 ng/ml) 15 minutes after stress (SPENCER 1980). Monitoring the concentrations of corticosteroids in the plasma of a boar before and after copulation revealed a sharp increase from 2 to 70 ng/ml (LIPTRAP & REASIDE 1977), but concentrations declined slowly to pretreatment levels several hours later. In addition, exposure of the same boar to an aggressive boar resulted in a sharp increase in the concentration of corticosteroids. Duration and maximum, however, were lower (15 to 47 ng/ml). In female prepuberal pigs (165 days of age) a corticosteroid response could be elicited by exposure to either a mature boar or an androgenized castrated male pig. Concentrations rose from about 7 ng/ml to levels of 20 ng/ml (PEARCE & HUGHES 1987).

Similarly, mating resulted in an increase of glucocorticoids in mature sows (BARNETT et al. 1982). In this study, however, the sow was mated in an unfamiliar pen. Glucocorticoid levels during mating were only 25% higher than those after tranportation of the animals to unfamiliar pens (7.5 vs. 5.6 ng/ml; pretreatment period: 2 ng/ml). These data show, that the release of glucocorticoids into the circulatory system is a very reliable stress indicator, which rapidly reacts even to moderate stressors (unfamiliar pen). After cessation of stress, however, glucocorticoids remain elevated for prolonged periods.

Another limiting factor is the high variation of endogenous levels in undisturbed animals due to an episodic secretion and circadian rhythms as shown in bulls (THUN et al. 1981).

Catecholamines

In contrast to the hormonal stimulation of the adrenal cortex, the adrenal medulla which releases adrenaline and noradrealine is under control of the sympathetic nervous system. Perhaps the most powerful stimulus for the release of catecholamines into the periphal plasma is physical stress such as maximal aerobic exercise as shown in humans (FLEG et al. 1985). During resting the plasma levels of adrenaline (A), noradrenaline (NA) and dopamine (DO) are very low in the species examined so far, as shown in Table 1.

Considerable species differences in catecholamine concentrations are obvious. In all species, however, the concentrations of NA exceed those of A and DO. Influences on the catecholamine release in humans were studied in detail. The reaction is dependent on age: advanced age is associated with

Table 1. Basal plasma concentrations of catecholamines (pg/ml blood plasma; $x \pm$ SEM) in different species. According to BÜHLER et al. (1978).

Species	no. of subjects	adrenaline	noradrenaline	dopamine
man	11	64 ± 5	203 ± 10	98 ± 20
cat	5	73 ± 17	609 ± 119	267 ± 7
rabbit	4	166 ± 39	392 ± 33	216 ± 23
dog	6	204 ± 12	376 ± 37	173 ± 61
cow	4	56 ± 12	152 ± 12	91 ± 35
rat (SPF)	18	175 ± 30	509 ± 46	84 ± 9

higher plasma concentrations of noradrenaline both during rest and in response to stress (FLEG et al. 1985). During maximal exercise, both adrenaline and noradrenaline increase up to 10-fold in plasma when compared to concentrations during rest (ZIEGLER et al. 1976). They return to basal values immediately after cessation of stress, due to the extremely short half-life of the catecholamines (less than 20 sec) in the peripheral blood (FERREIRA & VANE, 1967).

In contrast, studies on stress and catecholamine release in domestic farm animals are scanty. The effects of cold and heat exposure were investigated in cattle (DAVIS et al. 1984, KATTI et al. 1987, OLSON et al. 1981) and pigs (BENSON et al. 1986). Further studies were carried out in pigs by analyzing the influence of electroshocks (KEMPER et al. 1978) and noise (KEMPER et al. 1976). In pigs kept in confinement plasma catecholamine concentrations of about 400 pg/ml (KEMPER et al. 1976,1978) and noradrenaline concentrations of about 440 pg/ml were measured (BENSON et al. 1986). Studies on the influence of different types of physical stress on catecholamine release in the pig, however, are missing.

Measurement of catecholamines

Different methods to measure catecholamines are described in the literature. After separation from other blood constituents, catecholamines are converted into derivatives which lead to characteristic fluorescence spectra, so that they can be quantified with a high sensitivity and specifity. A very sensitive method was developed by DA PRADA and ZÜRCHER (1976) using an radioenzymatic labelling. Catecholamines were converted to their O-methylated derivatives by catechol-O-methyltransferase in the presence of tritiated S-adenosyl-methionine.

Alternatively, catecholamines were extracted from plasma with organic solvents and analysed by high-performance liquid chromatography (SMEDES et al. 1982). The catecholamines were separated on a hydrophobic column and quantified by oxidation on a glassy carbon electrode of an electrochemical detector. This method was adapted for our own study on catecholamine release in response to different stress conditions. In our assay system the lower limit of sensitivity was 80 pg/ml for adrenaline and 50 pg/ml for noradrenaline. For both catecholamines the interassay variation was 13% and the intraassay variation 7%. Due to the higher detection limit and the lower biological concentrations in case of adrenaline, we focussed our interest on the changes of the noradrenaline concentrations.

Stress mediated noradrenaline release in pigs

Six male and four female mature pigs of the German Landrace were used in this study. To ensure frequent blood collection without an additional stress, all animals were fitted with indwelling jugular vein catheters several weeks before the experiment. Samples were collected in 5 min intervals for a 20 min control period followed by the application of the stressor. During this application, blood was sampled at 2 min intervals and again at 5 min intervals after cessation of stress. Three different treatments were used: 1) transport of boars: four animals were loaded onto a lorry where they remained for 10 min; thereafter they were brought back to their familiar crates; 2) mating: catecholamines were measured in boars and sows; 3) dummy mounting: blood samples were collected from four boars during semen collection.

Loading

The profile of noradrenaline concentrations obtained from one of the boars (No.1) during the loading procedure is given in Fig. 2. Concomitantly to the start of the loading procedure (indicated by the black bar) noradrenaline concentrations increased and reached a maximum of 5 ng/ml at the very moment when the boar was on the lorry, where he remained for an additional 10 min. This period of rest is reflected by a dramatic decrease of noradrenaline and low concentrations thereafter. Moving the boar to his crate is reflected by a moderate increase (42 min). This individual profile shows that noradrenaline reacts very sensitive to the acute treatment.

Mating

The effect of mating on plasma noradrenaline for one of the boars and the corresponding sow is shown in Fig. 3. The first contact between both individuals occurred at 27 min, followed by mating attempts. At 34 min successful mating started. The duration of the ejaculation was 8 min (indicated by the black bar). In the boar plasma concentrations of noradrenaline rised dramatically from basal values (0.3 ng/ml) to a maximum of 11 ng/ml, decreasing immediately after the ejaculation. In contrast to the

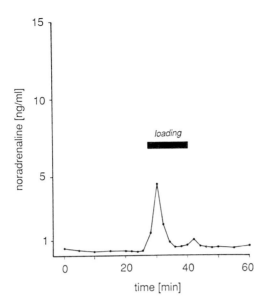

Fig. 2. Noradrenaline release of boar No.1 during the loading procedure. Loading started at 28 min and was finished at 30 min. The removal from the lorry was at 40 min.

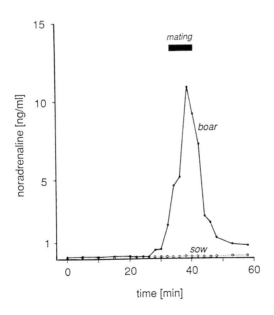

Fig. 3. Noradrenaline profiles of boar No.1 and sow No.1 during mating (27 min: first contact between both individuals; 27-34 min: mating attempts; 34-42 min: ejaculation indicated by the black bar).

boar, noradrenaline levels in the sow remained low, indicating that in this type of "stress" increasing noradrenaline concentrations are caused mainly by physical stress.

Dummy mounting and ejaculation

In comparison to mating, a similar profile was obtained for boar No.1 during semen collection on a dummy. There was again a tremendeous noradrenaline increase, reaching a maximum of 14 ng/ml which could be measured during dummy mounting and ejaculation, followed by a sudden decrease after cessation of stress (Fig.4).

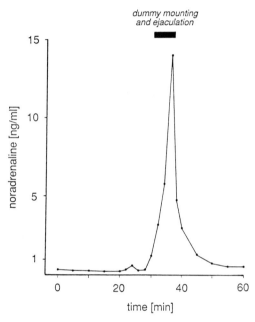

Fig. 4. *Noradrenaline profile of boar No.1 during dummy mounting and ejaculation. This period of activity is reflected by the horizontal bar.*

Comparison of stress conditions

The results obtained for the different treatments are summarized in Tab. 2. The data are expressed as maximal values with standard deviations for each treatment obtained from 4 animals each.

Compared to the basal concentrations (0.3 ng/ml) mating results in only a slight increase of noradrenaline in the four females investigated so far. In the boars, however, each treatment led to a

Table 2. *Maximal noradrenaline concentrations during the three treatments (ng/ml ± SD), n=4.*

	mean ± SD
mating *(male)*	6.98 ± 4.10
mating *(female)*	1.03 ± 1.15
dummy *(male)*	4.87 ± 6.12
transport *(male)*	3.07 ± 1.42

significant increase of noradrenaline in blood plasma which was highest during mating and less pronounced during dummy mounting and transport. However, it has to be noted that due to the limited number of animals used so far, a detailed graduation between the different treatments is not possible. This is mainly due to high variation within the different treatments, which are caused by differences in the magnitude of catecholamine release between individuals, as shown in Tab.3, where the maximal concentrations during the different treatments are compared for two boars.

Apparently, the reaction is dependent on the individual: boar No.1 reacted with high concentrations in all situations, whereas boar No.2, however, is only slightly affected by the same influences.

Table 3. Maximal noradrenaline concentrations obtained from boar No.1 and boar No.2 for each treatment (ng/ml).

	No.1	No.2
mating	10.9	1.8
dummy	14.0	0.9
transport	4.4	1.1

Conclusions

Noradrenaline is a very useful indicator for physical stress. In contrast, increasing adrenaline concentrations are detectable only if the exercise is severe (MUNRO & ROBINSON 1960). In humans changes in mood, especially when the subject shows fear, result in an increase of adrenaline secretion (CALLINGHAM 1975). Therefore, in studies with farm animals it is generally necessary to minimize the stress, which is connected with blood collection. In our study the stress was minimized by using indwelling jugular vein catheters and an appropriate control period before exposure to the stressor. Measurements of noradrenaline allow to monitor stress situations in detail, because it is released immediately into the periphal plasma. In contrast to glucocorticoids, where plasma concentrations remain elevated for a prolonged period, NA decreases immediately after cessation of stress. Between the individuals investigated, large differences in the response exist. It is likely that genetic differences are the explanation for the highly individual reaction.

References

BENSON, G.J.; LANGNER, P.H.; THURMON, J.C.; NELSON, D.R.; NEFF-DAVIS, M.S.; DAVIS, L.E.; TRANQUILLI, W.J.; GUSTAFSON, B.K. (1986): Plasma cortisol and norepinephrine concentrations in castrated male pigs maintained in pairs in outdoor pens and in a confinement finishing house: assessment of stress. Amer. J. Vet. Res. 47, 1071 - 1074.

BÜHLER, H.U.; DA PRADA, M.; HAEFELY, W.; PICOTTI, G.B. (1978): Plasma adrenaline, noradrenaline and dopamine in man and different animal species. J. Physiol. 276, 311 - 320.

CALLINGHAM, B.A. (1975): Catecholamines in blood. In: Handbook of Physiology. Section 7, Vol 6 (Editors: BLASCHKO, H., SAYERS, G., SMITH, A.D.). American Physiological Society, Washington. D.C. Pp. 427 - 445.

CANNON, W.B.; DE LA PAZ, D. (1911): Emotional stimulation of adrenal secretion. Amer. J. Physiol. 28, 64 - 70.

DÄMMRICH, K. (1987): Organ change and damage during stress. Morphological diagnosis. In: Biology of Stress in Farm Animals: An Integrative Approach (Editors: WIEPKEMA, P.R., VAN ADRICHEM, P.W.M). Martinus Nijhoff Publishers, Dordrecht - Boston - Lancaster. Pp. 71 - 81.

DA PRADA, M.; ZÜRCHER, G. (1976): Simultaneous radioenzymatic determination of plasma and tissue adrenaline, noradrenaline and dopamine within the femtomole range. Life Sci. 19, 1161 - 1174.

DAVIS, T.P.; JOHNSON, H.D.; GEHRKE, C.W. (1984): Effect of temperature stress on circulating biogenic amines in bovine. Comp. Biochem. Physiol. 79C, 369 - 373.

FERREIRA, S.H.; VANE, J.R. (1967): Half-lives of peptides and amines in the circulation. Nature 215, 1237 - 1240.

FLEG, J.L.; TZANKOFF, S.P.; LAKATTA, G. (1985). Age-related augmentation of plasma catecholamines during dynamic exercise in healthy males. J. Appl. Physiol. 59, 1033 - 1039.

KATTI, P.; JOHNSON, H.D.; GADDIS, R.R. (1987): Stress effects of environmental heat on plasma and milk catecholamine in dairy cattle. J. Dairy Sci. **70**, Suppl.1,123.

KEMPER, A.; WILDENHAHN, V.; LYHS, L. (1976): Die Einwirkung lang anhaltender Geräusche auf die Plasmakonzentration an Katecholaminen, Glukokortikosteroiden und PBJ bei Schweinen. Arch. Exp. Vet. Med., Leipzig **30**, 619 - 625.

KEMPER, A.; WILDENHAHN, V.; LYHS, L. (1978): Plasmakonzentration an Glukokortikosteroiden, Katecholaminen und proteingebundenem Jod (PBJ) nach Erregung durch elektrokutane Reizung beim Schwein. Arch. Exp. Vet. Med., Leipzig **32**, 879 - 885.

LENGERKEN, G.,v.; ALBRECHT, V.; SCHNEIDER, J. (1979): Einfluß der Halothan-Reaktivität beim Schwein auf das Verhalten biochemischer Kennwerte im Blut. Mh. Vet.-Med. **34**, 576 - 579.

LIPTRAP, R.M.; RAESIDE, J.I. (1978): A relationship between plasma concentrations of testosterone and corticosteroids during sexual and aggressive behaviour in the boar. J. Endocrinol. **76**, 75 - 85.

MUNRO, A.F.; ROBINSON, R. (1960): The catecholamine content of the peripheral plasma in human subjects with complete transverse lesions of the spinal cord. J. Physiol. **154**, 244 - 253.

OLIVERIA, A. (1987): Endocrine aspects of stress: Central and peripheral mechanisms. In: Biology of stress in farm animals: An integrative approach (Editors: WIEPKEMA, P. R., VAN ADRICHEM, P. W. M). Martinus Nijhoff Publishers, Dordrecht - Boston - Lancaster. Pp. 3 - 12.

OLSON, D.P.; RITTER, R.C.; PAPASIAN, C.J.; GUTENBERGER, S.(1981): Sympathoadrenal and adrenal hormonal responses of newborn calves to hypothermia. Can. J. Comp. Med. **45**, 321 - 326.

OLSON, J.D.; TRENKLE, A.(1973): Exposure of cattle to controlled subzero temperature : growth hormone, glucose, and free fatty acid concentrations in plasma. Amer. J. Vet. Res. **34**, 747 - 751.

PEARCE, G.P.; HUGHES, P.E. (1987): The influence of male contact on plasma cortisol concentrations in the prepuberal gilt. J. Reprod. Fert. **80**, 417 - 424.

PEREZ, E.; STEINMANN, C.; SCHOLL, E. (1988): Belastungsreaktion bei Ebern der Deutschen Landrasse und der Piétrainrasse während des Absamens am Phantom. Tierärztl. Prax. Suppl. **3**, 93 - 100.

REYNAERT, R.; MAROUS, S.; DE PAEPE, M.; PEETERS, G. (1976): Influences of stress, age and sex on serum growth hormone and free fatty acids level in cattle. Horm. Metab. Res. **8**, 109 -114.

SELYE, H. (1936): A syndrome produced by diverse nocuous agents. Nature **138**, 32.

SLEE, J.; HALLIDAY, R. (1968): Some effects of cold exposure, nutrition and experimental handling on serum free fatty-acid levels in sheep. Anim. Prod. **10**, 67 - 76.

SMEDES, F.; KRAAK, J.C.; POPPE, H. (1982): Simple and fast solvent extraction system for selective and quantitative isolation of adrenaline, noradrenaline an dopamine from plasma and urine. J. Chromatogr. **231**, 25 - 39.

SPENCER, G.S.G. (1980): Relationship between plasma somatomedin activity and levels of cortisol and free fatty acids following stress in pigs. J. Endocrinol. **84**, 109 - 114.

THUN, R.; EGGENBERGER, E.; ZEROBIN, K.; LÜSCHER, T.; VETTER, W. (1981): Twenty-four-hour secretory pattern of cortisol in the bull : Evidence of episodic secretion and circadian rhythm. Endocrinology **109**, 2208 - 2212.

Authors' address:
M. *Dehnhard and R. Claus, Universität Hohenheim, Institut für Tierhaltung und Tierzüchtung, POB 700562, D-7000 Stuttgart 70, F.R.G.*

A genetically-based model for divergent stress responses: behavioral, neurochemical and hormonal aspects

P. DRISCOLL, J. DEDEK, M. D'ANGIO, Y. CLAUSTRE and B. SCATTON

Stress and stressor

Whereas the "stressor" is the adverse environmental condition which impinges upon the complex adaptive apparatus of the animal, "stress" refers to the response of that apparatus (hormonal, neurochemical and behavioral), assumed according to the individual's ability to adjust to the situation. In other words, most stress-related behavioral and biochemical changes are dependent upon the possibility of the animal to gain control over the stressor, and not on its aversive or noxious nature *per se* (VOGEL 1985). Heat stress, for example, is the syndrome of hyperthermia in the animal, not the temperature of its ambient environment.

The "brain state" model

The hormonal, neurochemical and behavioral responses of the individual are all intimately interrelated, and are dependent upon the animal's genetic background and previous experience, both of which, in addition, influence the nature of the individual's perception of the stressor. According to the model proposed by DANTZER & MORMÈDE (1983), a threatening stimulus creates a "brain state" (at which stage the initial neurochemical changes occur, as determined by genetic factors and the animal's prior experience) and this, in turn, leads to the respective hormonal and behavioral responses. Our present state of knowledge, mainly as it pertains to recent studies concerned with hormonal feedback and brain receptors, permits us to expand this model to include subsequent neurochemical alterations which, presumably, account for the registration of "prior experience" in the animal's repertoire.

Genetic farm animal models

Most of the work done with farm animals along these lines has dealt with peripheral hormonal stress responses. Studies dealing with behavioral responses have been much less frequent, and adequate neurochemical experiments are almost nonexistent, especially ones taking vital genetic factors into consideration. Among the few exceptions have been the studies conducted in Spain by MUÑOZ-BLANCO & PORRAS CASTILLO (1987), in which females (!?) of the Spanish fighting bull and Friesian strains have been compared with regard to excitatory amino acid contents of several areas of the brain. These researchers have, of course, been dealing with the genetics and neurochemistry of aggression, which is not the subject of the present chapter. It may be noted, however, that although the experiments mentioned above were well executed, most other studies conducted on aggression to date have rather implicated monoaminergic and/or serotonergic mechanisms in that form of behavior (reviews on the subject include VALZELLI (1984), BELL & HEPPER (1987), VAN PRAAG et al. (1987). In another of the "exceptions", DRAPER and coworkers (1984) measured dopamine (DA), epinephrine and norepinephrine (NE) levels in the dorsal striatum of stress-susceptible and stress-resistant pigs which had been killed by electrical stunning. The omission of additional brain regions, as well as the lack of measurement of the metabolites of DA and NE, however, certainly limits the value of such experiments, as we will see.

DANTZER & MORMÈDE (1983), in addition to reviewing some genetic selection studies with chickens, also described hormonal experiments in which they compared two breeds of pigs. In a "chain pulling" study, they additionally showed that pigs which could behave, or did so, had lower corticosteroid levels at the end of a session than those which could not, or did not, pull the chain, this act being construed as a means of "dissipating anxiety". Considering that the most important farm animal stressors are probably the intensive methods of production used at present (i.e. artificial environment), and forced transport, it is remarkable that so few behavioral and neurochemical studies are being (have been) conducted, especially on pigs, which are the most common victims of such procedures. Much work on stress responses has been done with rodents, however, and there are, fortunately, enough similarities between the two to enable certain parallels to be drawn.

Like rodents, pigs are known to exhibit long periods of inactivity during the day, usually huddled, during these periods, in groups. They are also more active at night, showing a well-developed exploratory drive, rooting, etc. It is the latter drive, combined with the factors of over-crowding and cement floors (which make rooting impossible), which may be the underlying causes of tail biting. Pigs also show a clear-cut social hierarchy, as do rodents, and an inefficient thermoregulation. In the light of the previously-mentioned scarcity of adequate farm animal data, it will be the purpose of this chapter to present a genetically-based rodent model of divergent stress responses, with an emphasis being placed upon certain behavioral, neurochemical and hormonal aspects of the same.

A genetic rat model

Introduction

The Swiss lines of Roman high- and low-avoidance rats (RHA/Verh and RLA/Verh), respectively, are selected and bred for the rapid *versus* non-acquisition of two-way, active avoidance behavior. Numerous behavioral and physiological studies (for a review see DRISCOLL & BÄTTIG 1982) have indicated that this difference in avoidance acquisition is not due to "learning ability" but due, primarily, to emotional factors. RLA/Verh rats, which commonly "freeze" in the shuttlebox situation (rather than avoid or escape the shock), are considered to be more anxious than RHA/Verh rats. In addition to showing freezing behavior in shock situations, RLA/Verh rats, as compared to RHA/Verh rats, react negatively to repeated handling and injections (indicated by frequent defecation, urination, "tensing up" and squealing), show greater plasma corticosterone, ACTH and prolactin responses after, and higher defecation scores and lower activity levels during, various novel environment situations (GENTSCH et al. 1982, DRISCOLL & BÄTTIG 1982), and even show an augmentation of heart rate which is greater in magnitude and duration than that of RHA/Verh rats, in various stressful situations (D'ANGIO et al. 1988).

Important differences have been found between RHA/Verh and RLA/Verh rats in almost all neurochemical pathways which have been studied to date, e.g. in the serotonergic, peptidergic and

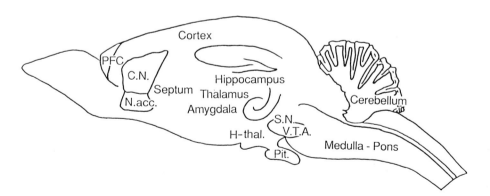

Fig. 1. Schematic representation of the rat brain. C.N., caudate nucleus; H-thal., hypothalamus; N.acc., nucleus accumbens; PFC, prefrontal cortex; Pit., pituitary; S.N., substantia nigra; V.T.A., ventral tegmental area.

cholinergic systems, in benzodiazepine-, imipramine- and opioid-receptor experiments, etc. However, this chapter will concentrate mainly on the nigrostriatal and mesocortical dopaminergic pathways, for both of which some new data will also be presented. Mention will be made of other systems only within the framework of proposing a concept of neurochemical and hormonal interactions which may regulate motor behavior and play a role in the direction of determining emotional and behavioral response to stressful situations. In this regard, reference will frequently be made to several special brain areas. Therefore, a simplified, schematic drawing of the rat brain is presented in Fig. 1, including the relative locations of all of the structures which will be referred to in the text.

Dopamine and Locomotion

Most DA-containing neurons are located in the substantia nigra (S.N.), and ventral tegmental area (V.T.A.) of the midbrain. The neurons from the S.N. project mainly to the caudate nucleus (C.N.), which is the dorsal striatum (or simply "striatum" in much of the earlier literature). This area also receives massive input from all major sensory areas of the neocortex, and sends efferents to important motor nuclei. The neurons from the V.T.A. project mainly to the nucleus accumbens and olfactory tubercles (ventral striatum), and lateral septum, all of which is known as the mesolimbic system, and to the prefrontal cortex (=mesocortical projection). The mesolimbic system, in particular, is also important in locomotion. The dorsal striatum has been much more extensively studied in RHA/Verh and RLA/Verh rats than has the ventral striatum, and our discussion of motor activity will, therefore, be mainly restricted to the nigrostriatal projection. FREED & YAMAMOTO (1985), however, have competently described the connections between changes in DA metabolism in both the C.N. and nucleus accumbens, and various types of movement in rats.

The experiments described in Table 1 indicate a higher basal turnover rate for striatal (C.N.) dopamine in RHA/Verh rats than for RLA/Verh rats. This was already apparent from observation of the logarithmic slopes showing the disappearance of the DA metabolite, dihydroxyphenylacetic acid (DOPAC), after monoamine oxidase inhibition by pargylene, as well as from measurements made on another DA metabolite, homovanillic acid (HVA), performed at the same time in three of the four experiments. On the other hand, no significant differences were found between the rat lines in basal, hypothalamic DA turnover. To obtain the data shown on Table 1, the levels of DA and DOPAC were assayed spectrofluorometrically. Adult males from each rat line were treated with an i.p. injection of pargylene (75 mg/kg b.w.) and kept afterwards in an undisturbed isolation for 0, 10, 20 and 30 min,

Table 1. *Basal, non-stressed, turnover rates of DA in the paired striata and hypothalami of RHA/Verh and RLA/Verh rats, as determined by the measurement of DOPAC disappearance after pargylene treatment. (19 animals for each of the four time periods for each line).*

	$t_{1/2}^{a}$ (min)	k^{a} (h^{-1})	turnover[b] ($\mu g\ h^{-1}/g$ tissue)
Striatum			
RHA/Verh	15.4	2.70	1.21
RLA/Verh	20.0	2.07	0.83
Hypothalamus			
RHA/Verh	50.0	0.83	0.22
RLA/Verh	47.0	0.87	0.21

[a]$t_{1/2}$, *half-life; k, fractional rate constant.*

[b]*Turnover rate computed as $k_{1/2} \times$ extrapolated tissue concentration ($\mu g/g$ tissue) at time 0 (cf. DRISCOLL et al. 1980).*

respectively. All treatments were counterbalanced and conducted during the middle six hours of the lighted part of the 12:12 hours cycle. The rats were decapitated following their respective isolation periods, and the brains were rapidly removed and dissected, with all areas of interest being immediately frozen on dry ice and stored at -70°C until they were assayed.

These results, when combined with those of other studies, which have shown that higher levels of the neuropeptide, substance P (SP), are consistently found in the C.N. and S.N. of RHA/Verh rats, as compared to RLA/Verh rats (DRISCOLL et al. 1988), suggest that the nigrostriatal DA system may be largely responsible for the differences in locomotor activity generally seen between the two lines of rats (BÄTTIG et al. 1976, DRISCOLL & BÄTTIG 1982, DRISCOLL 1986). Not only have numerous experiments supported the important role of the nigrostriatal system in motor activity, by associating increased DA release and metabolism with increased locomotion (e.g. O'NEILL & FILLENZ 1985, SPECIALE et al. 1986), but several have also shown that SP in the S.N. and/or C.N. increased the release of striatal DA while increasing locomotion and rearing behavior, i.e. both horizontal and vertical activity (KELLEY et al. 1985, HERRERA-MARSCHITZ et al. 1986, REID et al. 1988).

This action of SP in the nigrostriatal system is not unlike that of amphetamine. Although it is not the purpose of this chapter to consider pharmacological data in any detail, an exception might be made for amphetamine, as it is the only treatment of any kind to date which has been capable of (temporarily, at least) converting RLA/Verh rats into "high avoiders". It accomplished this by greatly stimulating the locomotor activity of that rat line, rather than by having any real (or permanent) effect on "learning" or emotionality (DRISCOLL 1986). These findings are not surprising, as repeated injections of amphetamine have been shown to increase DA release and affect DA turnover in the entire striatum, while increasing locomotor activity at the same time (SEGAL & KUCZENSKI 1987). The same authors also emphasized the importance of "individual differences in responsiveness" to amphetamine, which had been extremely well-illustrated by the contrasting behavioral results seen with RHA/Verh and RLA/Verh rats in the previously-mentioned study (DRISCOLL 1986). Other interesting findings by SEGAL & KUCZENSKI (1987) concerned the effects of amphetamine (direct or indirect?) on the serotonergic system in the striatum, frontal cortex and hippocampus, where an increase in serotonin (5-hydroxytryptamine; 5-HT) and a reduction in the metabolite of 5-HT, 5-hydroxyindolacetic acid (5-HIAA) were noted, as well as a lack of effect of amphetamine on the noradrenergic system.

To conclude this discussion of the striatal dopaminergic system and locomotion, another study which was undertaken with RHA/Verh and RLA/Verh rats should be considered. Immediately before sacrifice and dissection, groups of naive, adult rats from each line were exposed either to normal housing conditions (cage control), or to a 30 min exploratory session in the shuttlebox (without shock), or to 40 inescapable, scrambled shocks during the 30 min, while confined to one side of the shuttlebox, or to 40 avoidable (or escapable, when need be) shocks under two-way avoidance conditions, also during 30 min, following by one week a previous acquisition session so that the rats (RHA/Verh only, of course) already showed a high level of avoidance (DRISCOLL et al. 1983). In that study, highly significant results were found in regard to the serotonergic system in several brain regions, whereas the results for the dopaminergic system were considered not to be particularly meaningful. It should be pointed out, however, that for reasons of comparison of the DA, DOPAC and HVA data with the data presented for the other neurotransmitter systems, the "cage control" condition, which had only been applied during the DA part of the study, had been omitted from the results, leaving only comparisons with the "shuttlebox control" condition. The present Table 2 corrects this omission, showing the striatal data for DA, DOPAC and HVA in their entirety.

Looking at the results for DOPAC, (which can be advantageously compared with experiments to be discussed later, in which this DA metabolite was also measured - see next section), it can be seen that i) an increase in DOPAC (i.e. DA metabolism) occurred in RLA/Verh rats which were moving around (i.e. exploring the shuttlebox without being shocked), ii) this increase disappeared during the shock-stressor condition, at which time the RLA/Verh rats were "freezing" (or, more or less, completely motionless), iii) the biggest increase in DOPAC for RHA/Verh rats, on the other hand, occurred during the shock-stressor condition, when the animals were most active, i.e. running around trying to escape from the shock, and iv) a significant increase in DA metabolism was also found during the avoidance sessions, at which time the RHA/Verh rats were actively avoiding the shocks at regular intervals and were otherwise rather active. These results confirm the view that enhanced locomotor activity is associated with an increased striatal DA turnover. It may also be worth mentioning (perhaps comparing these results and methods, at the same time, with the experiments with pigs of DRAPER and coworkers (1984), which were mentioned earlier), that measurements of metabolites are more important than those of mere DA levels in ascertaining the functional state of the dopaminergic system and, also, that we must look elsewhere in the brain if we want to learn more about coping with stressors, and not just about motor responses (see WESTERINK 1985, BEAL & MARTIN 1985, O'NEILL & FILLENZ 1985, CLAUSTRE et al. 1986, SPECIALE et al. 1986).

Dopamine and coping with stressors

In contrast to the prevailing view held at several laboratories, which have been portraying the prefrontal cortex of the rat as a sort of adrenal gland in the brain, reflecting the emotional state of the animal by releasing DA in response to a stressor (much as the adrenal cortex releases corticosterone), an increasing number of researchers have been ascribing functions to this region more in keeping with what has been known for primates for several years. Some of the functions which have been suggested for the prefrontal cortex in rats now include, for example, i) serving as the rat's "frontal eye field", involved in head and eye orientation and, therefore, in attention (NEAFSEY et al. 1986), ii) being uniquely responsive to light and sound (MIRMIRAN et al. 1986), iii) being involved in higher-order functioning and the temporal structuring of information (DOAR et al. 1987, KESNER & HOLBROOK 1987, GROENEWEGEN 1988), iv) "searching for an adequate behavioral strategy" (MOGENSEN & JORGENSEN 1987) and v) providing an excitatory influence on motor control (MORENCY et al. 1987). The results of experiments with RHA/Verh and RLA/Verh rats which have been concerned with this subject may be said to be, happily, more or less in agreement with all of these (recent) suggestions. In those studies, the effects of a variety of stressful environmental situations on DA metabolism in the prefrontal cortex (i.e. the level of extracellular DOPAC, as assessed by *in vivo* voltammetry with carbon fiber electrodes), were compared in both lines of rats. All of the stressors (novel environment, tail pinch, a loud noise or immobilization) resulted in an increased DA metabolism in the less emotional RHA/Verh rats but not in the more emotional RLA/Verh rats, while provoking at the same time an increased heart rate, increased defecation and freezing, etc., in the latter (D'ANGIO et al. 1988). It was therefore concluded that the increase in cortical DA metabolism was not connected to an "emotional reaction" caused by the aversive nature of the stressor, but rather to an heightened attention of the animal or activation of cognitive processes in an attempt to cope with it.

This conclusion was supported by two experiments in which tissue DOPAC in the prefrontal cortex was measured by HPLC, immediately following all of the conditions used in the study which was previously summarized in Table 2. The results of these experiments, pooled together and expressed as percent from baseline, are shown in Table 3. In comparison to the striatal results, it can be seen that endogenous DOPAC levels increased significantly in RHA/Verh rats exposed to inescapable footshocks, but not in those RHA/Verh rats which were allowed to avoid (about 80-85 percent of the time) or to terminate (by escaping the other 15-20 percent of the time) the shocks. The latter rats, of course, had been allowed to cope with the stressor. It should also be noted that DOPAC's "locomotor effects", observed for both lines of rats in the caudate nuclei, were not in evidence in the prefrontal cortex. Our results are in agreement with those of others who have measured DA metabolism in the striatum (SPECIALE et al. 1986) and/or prefrontal cortex (FEKETE et al. 1981, HERMAN et al. 1982, GLOWINSKI et al. 1984, MILLER et al. 1984, CLAUSTRE et al. 1986, SPECIALE et al. 1986) after footshock, several of whom have expressed doubts, in one form or another, that the prefrontal dopaminergic system is directly involved in emotionality, or anxiety, *per se* (FEKETE et al. 1981, HERMAN et al. 1982, CLAUSTRE et al. 1986). It should be mentioned at this point that some of these studies also reported no dorsal striatal changes in DA metabolism following footshock. With very few exceptions, however (e.g. SPECIALE et al. 1986), no mention was made of activity levels of the rats during application of the footshock, and it is quite likely that many of the rats involved, (especially taking into account the methods and shock levels most often used), "froze" upon repeated shock presentations.

Table 2. DA, DOPAC and HVA levels in the caudate nuclei of RHA/Verh and RLA/Verh rats after different locomotor and/or stressor conditions.

	DA[a]	DOPAC[a]	HVA[a]
RHA/Verh			
cage control	5.97±0.43(A)	0.63±0.03(B)	0.21±0.02(C)
shuttlebox control	6.48±0.48	0.71±0.05	0.27±0.03
inescapable shock	6.84±0.50	0.82±0.06(b)	0.33±0.04(c)
acquired avoidance	7.59±0.42(a)	0.77±0.04(b)	0.31±0.03(c)
RLA/Verh			
cage control	7.62±0.29(a)	0.64±0.02(B)	0.23±0.02
shuttlebox control	7.48±0.35	0.78±0.07(b)	0.31±0.03
inescapable shock	7.11±0.52	0.70±0.07	0.31±0.03

[a]*Expressed as* µg/g *tissue ± standard error of mean. 7 animals per group. Difference significant at the probability level 0.05 was assessed between values A/a, B/b, and C/c (symbols in parentheses).*

That a significant increase in prefrontal DOPAC is considered to reflect a heightened attention, or activation of cognitive processes, is not surprising, in view of the varied connections which have been demonstrated between the prefrontal cortex and other brain structures known to be involved in emotionality, "coping" and/or locomotion. Mesohippocampal and direct fronto-striatal (and -brainstem) connections have been known to exist for several years (e.g. SCATTON et al. 1980, BENINGER 1983) and, more recently, direct connections from the prefrontal cortex to the hippocampus and thalamus have also been revealed (KILPATRICK & PHILLIPSON 1986, FERINO et al. 1987, GROENEWEGEN 1988). Another structure with mesolimbic connections, which certainly plays a vital role in several aspects of animal behavior, is the amygdala (BEAULIEU et al. 1986, 1987). Indeed, it has already been suggested that the amygdala (with special reference to its central nucleus) is probably very important in the differences in emotionality found between RHA/Verh and RLA/Verh rats (HENKE 1988). As few neurochemical studies have been conducted to date in this structure in the Roman lines, it would be premature to include further consideration of it, at this time. There is no doubt, however, that this structure will have to be taken into account in future versions of this behavioral model.

An "interim" model

Any model contemplating the regulation of stressor-induced behaviors, mood, neuroendocrine and cardiovascular functions, etc., must at least take into consideration the diffusely projecting stress-responsive brain stem systems, including S.N., V.T.A. and hypothalamic catecholaminergic and serotonergic neurons, as well as the septo-hippocampal cholinergic system, all of which are under the influence of other neuromodulators, and of circulating glucocorticoid hormones (GRAY 1982, HÄRFSTRAND et al. 1986). As a prelude to constructing such an "interim" model, therefore, additional information must be provided regarding these systems in the Roman rat lines, at least from a neurochemical standpoint. First of all, based on basal measurements for cholineacetyltransferase (Chat) activity in the hippocampus, it may be rationalized that RLA/Verh rats appear to have a more

Table 3. DOPAC levels in the dorso-medial, prefrontal cortices of RHA/Verh and RLA/Verh rats, after different locomotor and/or stressor conditions, expressed as percent from baseline. Number of animals per group: 2 repetitions, 7 animals each. (See text for details).

		DOPAC[a]
RHA/Verh	cage control	100
	shuttlebox control	123
	inescapable shock	168[b]
	acquired avoidance	130
RLA/Verh	cage control	105
	shuttlebox control	116
	inescapable shock	137

[a]Per cent baseline (cage control RHA/Verh).
[b]Value significantly different from baseline ($p < 0.05$).

active hippocampal cholinergic system than do RHA/Verh rats, i.e. a higher production and turnover rate of acetylcholine (DRISCOLL et al. 1987). Secondly, RHA/Verh rats have been found to have a much higher basal turnover rate of serotonin (5-HT) in the hypothalamus (DRISCOLL 1988) and midbrain (DRISCOLL et al. 1980) than do RLA/Verh rats. On the other hand, RLA/Verh rats have shown an increased metabolism of 5-HT after acute footshock presentations, whereas RHA/Verh rats have shown an attenuated 5-HT metabolism and/or synthesis rate after the same treatment, and after avoidance learning (DRISCOLL et al. 1983). RLA/Verh rats have also been found to be more sensitive to the aversive effects of lateral hypothalamic self-stimulation (LIPP 1979).

On the neuroendocrine side, greater hormonal responses (corticosterone, ACTH, etc.) in RLA/Verh rats, after various novel environment situations, have been described earlier in this chapter. It should be further mentioned that, in more recent studies, RHA/Verh pituitary glands have been found (through the measurement of ACTH secretion) to be less responsive than RLA/Verh pituitary glands to corticotropin releasing factor (CRF), both *in vivo* and *in vitro* (WALKER et al. 1989). In those experiments, the elevated, stressor-induced ACTH output in the RLA/Verh line was associated with an enhanced pituitary sensitivity to CRF and, possibly, with a "diminished corticosterone-inhibitory-feedback action on CRF and ACTH secretion".

As far as the established, behavioral and hormonal responses to footshock of RLA/Verh rats are concerned, recent experiments conducted at other laboratories unequivocally support the proposed model. KALIN et al. (1988) have shown that brain CRF systems mediate stressor-induced freezing behavior, an index of a rat's level of fear, probably not doing so by altering pain sensitivity. Others, e.g. FANSELOW & HELMSTETTER (1988), have stated that freezing and analgesia both seem to be mediated by a common fearlike process. (At least all seem to agree on the role of CRF in increased fearfulness and shock-induced freezing!). In any case, it is known that CRF from the paraventricular nuclei of the hypothalamus plays a prominent role in regulating the stressor-induced release of ACTH from anterior pituitary corticotrophic cells, and KALIN et al. (1988) have been able to increase shock-induced freezing by administering CRF intracerebroventricularly. Furthermore, FELDMAN et al. (1987) have suggested that paraventricular 5-HT mediates the adrenocortical (plasma corticosterone) response following afferent neural stimuli by controlling the release of CRF, bringing to mind the increased activation of the hypothalamic serotonergic system in (only) RLA/Verh rats upon exposure to footshocks.

Having thus far worked on the "model" from both ends, so to speak (i.e. the mesocortical, and hypothalamo-pituitary-adrenal, aspects), a most logical meeting place for the two might be the septo-hippocampal system. Although this region has been the center point of entire hypotheses regarding anxiety (e.g. GRAY 1982), and certainly plays a role in the adaptive response to various stressors, its exact role is not yet clear. In addition to studies which have revealed prominent correlations between avoidance behavior and the extent of the infrapyramidal mossy fiber projection (SCHWEGLER & LIPP 1983, LIPP et al. 1988), other genetically-oriented studies seem to point in the direction of the (septo-)hippocampal cholinergic system mediating a suppressive behavioral response to environmental stressors, perhaps even suppressing motor behavior in "stress-susceptible" rats without particularly affecting their avoidance capabilities (SCHMIDT et al. 1980, BLAKER et al. 1983, DRISCOLL et al. 1985). The latter experiment, performed at this laboratory, dealt with the effects of large, bilateral lesions in the dorsal hippocampi of RHA/Verh and RLA/Verh rats. It was found that non-trained, lesioned RHA/Verh rats were more active in the shuttlebox (i.e. showed more intertrial responses, or ITRs) than were non-trained controls, but that there was no difference in avoidance acquisition between the two groups. In addition, lesioned or non-lesioned, previously-trained RHA/Verh rats maintained their previous, high levels of avoidance performance. On the other hand, shuttlebox-naive, lesioned RLA/Verh rats had more ITRs and froze less than did their shuttlebox-naive controls, but avoidance acquisition occurred at the same (low) level in both groups. So the hyperactivity produced by disruption of the "suppressive behavioral response to environmental stressors" did not improve the performance of RLA/Verh rats in this test. At the same time, this study demonstrated that the high levels of activity normally shown by RHA/Verh rats are not *per se* responsible for their rapid acquisition of two-way avoidance.

However one wishes to interpret the behavioral aspects of septo-hippocampal function, the anatomical and neuro-hormonal reasons for this region's being the ideal "meeting place" certainly do exist (GILAD 1987). Direct DA innervation of the lateral septum is part of the mesolimbic dopaminergic system which, as we have also seen, is highly activated by environmental stressors. GILAD (1987) has proposed that the mesoseptal system exerts a tonic inhibitory action on the septo-hippocampal cholinergic system, with GABA perhaps providing the transneuronal transfer of information from DA terminals in the lateral septum to the parent cholinergic perikarya in the medial septum, via inhibitory GABA interneurons. In addition to serotonergic inputs to this system, which are probably also involved in the stress response, high blood levels of pituitary-adrenocortical hormones (as seen during the stress response of RLA/Verh rats, for example) are capable of indirectly, or directly, activating the hippocampal cholinergic terminals, by acting at corticosterone receptors on dopaminergic and/or cholinergic neurons, to affect hippocampus-associated functions such as cognition and mood (DE KLOET & REUL 1987, GILAD 1987).

For the time being, therefore, it may be hypothesized that stressors (e.g. footshock) preferentially stimulate dopaminergic neurons in the S.N. and V.T.A. of RHA/Verh rats, and the hypothalamic serotonergic system of RLA/Verh rats, leading to an immediate increase in motor activity of RHA/Verh rats and a freezing (inactivity) response in RLA/Verh rats through, respectively, stimulation of striatal activity in the former and an increased secretion of hypothalamic CRF and pituitary ACTH in the latter. At the same time, the preferential stimulation of mesolimbocortical activity in the RHA/Verh rats leads to heightened attention and an activation of cognitive processes in an attempt to cope with the stressor in that line (perhaps both prefrontal and hippocampal in origin and, in any case, lending further support to the increased-motor-activity solution for the particular problem), whereas the ensuing, preferential stimulation of the adrenal cortex in RLA/Verh rats results in high levels of circulating corticosterone which exerts its feedback effects primarily in the hippocampus, where it is capable of mobilizing the (behavior suppressing) cholinergic terminals.

The simplified ideas put forward in the previous paragraph certainly represent only a part of what, at some later date, will be the final picture, but it is reasonable to assume that they are probably at least a step in the direction toward an understanding of the divergent stress responses to be found within animal species. A definitive, genetic analysis awaits a clarification of the roles of other endogenous systems and pathways (e.g. SP, GABA, the amygdala, etc. - all of which were briefly mentioned here - among others), as they pertain to the modulation of, and interactions with, the proposed neurotransmitter and hormonal network which has been discussed in this chapter.

References

BÄTTIG, K.; DRISCOLL, P.; SCHLATTER, J.; USTER, H.J. (1976): Effects of nicotine on the exploratory locomotion patterns of female Roman high- and low-avoidance rats. Pharmacol. Biochem. Behav. **4**, 435-439.

BEAL, M.F.; MARTIN, J.B. (1985): Topographical dopamine and serotonin distribution and turnover in rat striatum. Brain Res. **358**, 10-15.

BEAULIEU, S.; DI PAOLO, T.; BARDEN, N. (1986): Control of ACTH secretion by the central nucleus of the amygdala: implication of the serotoninergic system and its relevance to the glucocorticoid delayed negative feedback mechanism. Neuroendocrinology **44**, 247-254.

BEAULIEU, S.; DI PAOLO, T.; CÔTÉ, J.; BARDEN, N. (1987): Participation of the central amygdaloid nucleus in the response of adrenocorticotropin secretion to immobilization stress: opposing roles of the noradrenergic and dopaminergic systems. Neuroendocrinology **45**, 37-46

BELL, R.; HEPPER, P.G. (1987): Catecholamines and aggression in animals. Behav. Brain Res. **23**, 1-21.

BENINGER, R.J. (1983): The role of dopamine in locomotor activity and learning. Brain Res. Rev. **6**, 173-196.

BLAKER, W.D.; CHENEY, D.L.; STOFF, D.M. (1983): Interstrain comparison of avoidance behavior and neurochemical parameters of brain cholinergic function. Pharmacol. Biochem. Behav. **18**, 189-193.

CLAUSTRE, Y.; RIVY, J.P.; DENNIS, T.; SCATTON, B. (1986): Pharmacological studies on stress-induced increase in frontal cortical dopamine metabolism in the rat. J. Pharmacol. Exp. Ther. **238**, 693-700.

D'ANGIO, M.; SERRANO, A.; DRISCOLL, P.; SCATTON, B. (1988): Stressful environmental stimuli increase extracellular DOPAC levels in the prefrontal cortex of hypoemotional (Roman high-avoidance) but not hyperemotional (Roman low-avoidance) rats. An in vivo voltammetric study. Brain Res. **451**, 237-247.

DANTZER, R.; MORMÈDE, P. (1983): Stress in farm animals: a need for reevaluation. J. Anim. Sci. **57**, 6-18.

DE KLOET, E.R.; REUL, J.M.H.M. (1987): Feedback action and tonic influence of corticosteroids on brain function: a concept arising from the heterogeneity of brain receptor systems. Psychoneuroendocrinology **12**, 83-105.

DOAR, B.; FINGER, S.; ALMLI, C.R. (1987): Tactile-visual acquisition and reversal learning deficits in rats with prefrontal cortical lesions. Exp. Brain Res. **66**, 432-434.

DRAPER, D.D.; ROTHSCHILD, M.F.; BEITZ, D.C.; CHRISTIAN, L.L. (1984): Age- and genotype-dependent differences in catecholamine concentrations in the porcine caudate nucleus. Exp. Gerontol. **19**, 377-381.

DRISCOLL, P. (1986): Roman high- and low-avoidance rats: present status of the Swiss sublines, RHA/Verh and RLA/Verh, and effects of amphetamine on shuttle-box performance. Behav. Genet. **16**, 355-364.

DRISCOLL, P. (1988): Hypothalamic serotonin turnover in rat lines selectively bred for differences in active, two-way avoidance behavior. Adv. Biosci. **70**, 55-58.

DRISCOLL, P.; BÄTTIG, K. (1982): Behavioral, emotional and neurochemical profiles of rats selected for extreme differences in active, two-way avoidance performance. In: Genetics of the Brain (Editor: LIEBLICH I.), Elsevier Biomedical Press, Amsterdam, Pp. 95-123.

DRISCOLL, P.; DEDEK, J.; MARTIN, J.R.; BÄTTIG, K. (1980): Regional 5-HT analysis in Roman high- and low-avoidance rats following MAO inhibition. Eur. J. Pharmacol. **68**, 373-376.

DRISCOLL, P.; DEDEK, J.; MARTIN, J.R.; ZIVKOVIC, B. (1983): Two-way avoidance and acute shock stress induced alterations of regional noradrenergic, dopaminergic and serotonergic activity in Roman high- and low-avoidance rats. Life Sci. **33**, 1719-1725.

DRISCOLL, P.; FITZGERALD, R.; BÄTTIG, K.; LIPP, H.P. (1985): Bilateral, dorsal hippocampal lesions do not alter the two-way avoidance performance of either Roman high-avoidance or Roman low-avoidance rats. Behav. Genet. **15**, 592.

DRISCOLL, P.; CLAUSTRE, Y.; FAGE, D.; SCATTON, B. (1987): Recent findings in central dopaminergic and cholinergic neurotransmission of Roman high- and low-avoidance rats. Behav. Brain Res. **26**, 213.

DRISCOLL, P.; CLAUSTRE, Y.; OBLIN, A.; DEDEK, J.; ZIVKOVIC, B.; SCATTON, B. (1988): Dopamine metabolism and substance-P levels within the nigro-striatal and meso-cortical projections of Roman high- and low-avoidance rats following increased locomotor activity, footshock stress and two-way active avoidance acquisition. Behav. Genet. **18**, 715.

FANSELOW, M.S.; HELMSTETTER, F.J. (1988): Conditioned analgesia, defensive freezing, and benzodiazepines. Behav. Neurosci. **102**, 233-243.

FEKETE, M.I.K.; SZENTENDREI, T.; KANYICSKA, B.; PALKOVITS, M. (1981): Effects of anxiolytic drugs on the catecholamine and DOPAC levels in brain cortical areas and on corticosterone and prolactin secretion in rats subjected to stress. Psychoneuroendocrinology **6**, 113-120.

FELDMAN, S.; CONFORTI, N.; MELAMED, E. (1987): Paraventricular nucleus serotonin mediates neurally stimulated adrenocortical secretion. Brain Res. Bull. **18**, 165-168.

FERINO, F.; THIERRY, A.M.; GLOWINSKI, J. (1987): Anatomical and electrophysiological evidence for a direct projection from Ammon's horn to the medial prefrontal cortex in the rat. Exp. Brain Res. **65**, 421-426.

FREED, C.R.; YAMAMOTO, B.K. (1985): Regional brain dopamine metabolism: a marker for the speed, direction, and posture of moving animals. Science **229**, 62-65.

GENTSCH, C.; LICHTSTEINER, M.; DRISCOLL, P.; FEER, H. (1982): Differential hormonal and physiological responses to stress in Roman high- and low-avoidance rats. Physiol. Behav. **28**, 259-263.

GILAD, G.M. (1987): The stress-induced response of the septo-hippocampal cholinergic system. A vectorial outcome of psychoneuroendocrinological interactions. Psychoneuroendocrinology **12**, 167-184.

GLOWINSKI, J.; TASSIN, J.P.; THIERRY, A.M. (1984): The mesocortical-prefrontal dopaminergic neurons. Trends Neurosci. **7**, 415-418.

GRAY, J.A. (1982): The Neuropsychology of Anxiety: An Enquiry into the Functions of the Septo-Hippocampal System. Oxford Science Publications, Oxford. Pp. 1-548.

GROENEWEGEN, H.J. (1988): Organization of the afferent connections of the mediodorsal thalamic nucleus in the rat, related to the mediodorsal-prefrontal topography. Neuroscience **24**, 379-431.

HÄRFSTRAND, A.; FUXE, K.; CINTRA, A.; AGNATI, L.F.; ZINI, I.; WIKSTRÖM, A.-C.; OKRET, S.; YU, Z.-Y.; GOLDSTEIN, M.; STEINBUSCH, H.; VERHOFSTAD, A.; GUSTAFSSON, J.-A. (1986): Glucocorticoid receptor immunoreactivity in monoaminergic neurons of rat brain. Proc. Natl. Acad. Sci. **83**, 9779-9783.

HENKE, P.G. (1988): Electrophysiological activity in the central nucleus of the amygdala: emotionality and stress ulcers in rats. Behav. Neurosci. **102**, 77-83.

HERMAN, J.P.; GUILLONNEAU, D.; DANTZER, R.; SCATTON, B.; SEMERDJIAN-ROUQUIER, L.; LE MOAL, M. (1982): Differential effects of inescapable footshocks on dopamine turnover in cortical and limbic areas of the rat. Life Sci. **30**, 2207-2214.

HERRERA-MARSCHITZ, M.; CHRISTENSSON-NYLANDER, I.; SHARP, T.; STAINES, W.; REID, M.; HÖKFELT, T.; TERENIUS, L.; UNGERSTEDT U. (1986): Striato-nigral dynorphin and substance P pathways in the rat. Exp. Brain Res. **64**, 193-207.

KALIN, N.H.; SHERMAN, J.E.; TAKAHASHI, L.K. (1988): Antagonism of endogenous CRH systems attenuates stress-induced freezing behavior in rats. Brain Res. **457**, 130-135.

KELLEY, A.E.; CADOR, M.; STINUS, L. (1985): Behavioral analysis of the effect of substance P injected into the ventral mesencephalon on investigatory and spontaneous motor behavior in the rat. Psychopharmacology **85**, 37-46.

KESNER, R.P.; HOLBROOK, T. (1987): Dissociation of item and order spatial memory in rats following medial prefrontal cortex lesions. Neuropsychologia **25**, 653-664.

KILPATRICK, I.C.; PHILLIPSON, O.T. (1986): Thalamic control of dopaminergic functions in the caudate-putamen of the rat - I. The influence of electrical stimulation of the parafascicular nucleus on dopamine utilization. Neuroscience **19**, 965-978.

LIPP, H.-P. (1979): Differential hypothalamic self-stimulation behavior in Roman high-avoidance and low-avoidance rats. Brain Res. Bull. **4**, 553-559.

LIPP, H.-P.; SCHWEGLER, H.; HEIMRICH, B.; DRISCOLL, P. (1988): Infrapyramidal mossy fibers and two-way avoidance learning: developmental modification of hippocampal circuitry and adult behavior of rats and mice. J. Neurosci. **8**, 1905-1921.

MILLER, J.D.; SPECIALE, S.G.; MCMILLEN, B.A.; GERMAN, D.C. (1984): Naloxone antagonism of stress-induced augmentation of frontal cortex dopamine metabolism. Eur. J. Pharmacol. **98**, 437-439.

MIRMIRAN, M.; BRENNER, E.; VAN GOOL, W.A. (1986): Visual and auditory evoked potentials in different areas of rat cerebral cortex. Neurosci. Letters **72**, 272-276.

MOGENSEN, J.; JORGENSEN, O.S. (1987): Protein changes in the rat's prefrontal and "inferotemporal" cortex after exposure to visual problems. Pharmacol. Biochem. Behav. **26**, 89-94.

MORENCY, M.A.; STEWART, R.J.; BENINGER, R.J. (1987): Circling behavior following unilateral microinjections of cocaine into the medial prefrontal cortex: dopaminergic or local anesthetic effect? J. Neurosci. **7**, 812-818.

MUÑOZ-BLANCO, J.; PORRAS CASTILLO, A. (1987): Changes in neurotransmitter amino acids content in several CNS areas from aggressive and non-aggressive bull strains. Physiol. Behav. **39**, 453-457.

NEAFSEY, E.J.; HURLEY-GIUS, K.M.; ARVANITIS, D. (1986): The topographical organization of neurons in the rat medial frontal, insular and olfactory cortex projecting to the solitary nucleus, olfactory bulb, periaqueductal gray and superior colliculus. Brain Res. **377**, 261-270.

O'NEILL, R.D.; FILLENZ, M. (1985): Simultaneous monitoring of dopamine release in rat frontal cortex, nucleus accumbens and striatum: effect of drugs, circadian changes and correlations with motor activity. Neuroscience **16**, 49-55.

REID, M.; HERRERA-MARSCHITZ, M.; HÖKFELT, T.; TERENIUS, L.; UNGERSTEDT, U. (1988): Differential modulation of striatal dopamine release by intranigral injection of GABA, dynorphin A and substance P. Eur. J. Pharmacol. **147**, 411-420.

SCATTON, B.; SIMON, H.; LE MOAL, M.; BISCHOFF, S. (1980): Origin of dopaminergic innervation of the rat hippocampal formation. Neurosci. Letters **18**, 125-131.

SCATTON, B.; D'ANGIO; M.; DRISCOLL, P.; SERRANO, A. (1988): An *in vivo* voltammetric study of the response of mesocortical and mesoaccumbens dopaminergic neurons to environmental stimuli in strains of rats with differing levels of emotionality. Ann. N.Y. Acad. Sci. **537**, 124-137.

SCHMIDT, D.E.; COOPER, D.O.; BARRETT, R.J. (1980): Strain specific alterations in hippocampal cholinergic function following acute footshock. Pharmacol. Biochem. Behav. **12**, 277-280.

SCHWEGLER, H.; LIPP, H.-P. (1983): Hereditary covariations of neuronal circuitry and behavior: correlations between the proportions of hippocampal synaptic fields in the regio inferior and two-way avoidance in mice and rats. Behav. Brain Res. **7**, 1-38.

SEGAL, D.S.; KUCZENSKI, R. (1987): Individual differences in responsiveness to single and repeated amphetamine administration: behavioral characteristics and neurochemical correlates. J. Pharmacol. Exp. Ther. **242**, 917-926.

SPECIALE, S.G.; MILLER, J.D.; MCMILLEN, B.A.; GERMAN, D.C. (1986): Activation of specific central dopamine pathways: locomotion and footshock. Brain Res. Bull. **16**, 33-38.

VALZELLI, L. (1984): Reflections on experimental and human pathology of aggression. Prog. Neuro-Psychopharmacol. **8**, 311-325.

VAN PRAAG, H.M.; KAHN, R.S.; ASNIS, G.M.; WETZLER, S.; BROWN, S.L.; BLEICH, A.; KORN, M.L. (1987): Denosologization of biological psychiatry (or) the specificity of 5-HT disturbances in psychiatric disorders. J. Affect. Disord. **13**, 1-8.

VOGEL, W.H. (1985): Coping, stress, stressors and health consequences. Neuropsychobiology **13**, 129-135.

WALKER, C.-D.; RIVEST, R.W.; MEANEY, M.J.; AUBERT, M.L. (1989): Differential activation of the pituitary-adrenocortical axis after stress in the rat: use of two genetically selected lines (Roman low- and high-avoidance rats) as a model. J. Endocrinol., in print.

WESTERINK, B.H.C. (1985): Sequence and significance of dopamine metabolism in the rat brain. Neurochem. Int. **7**, 221-227.

Authors' addresses:
P. Driscoll, Laboratorium für vergleichende Physiologie und Verhaltensbiologie, ETH Zürich, CH - 8092 Zürich, Switzerland;
M. D'Angio, Y. Claustre and B. Scatton, Synthelabo (LERS), F - 92220 Bagneux, France.;
J. Dedek, Glaxo IMB S.A., CH - 1211 Genève, Switzerland.

Disease models

Two unique animal models for the study of human metabolic bone diseases

J. Harmeyer and R. Kaune

Summary

Inherited diseases have been described in animals whose symptoms closely resemble those of distinct inherited disorders in humans. The animals which suffer from such diseases have played an important role in the study of pathomechanisms and etiology of the corresponding human diseases. The diseased animals are also of importance for development and testing of effective treatments. X-linked hypophosphatemic rickets (XLH) is an inborn error of phosphate metabolism which occurs in humans and in mice. XLH is the most common form of human vitamin D-resistant rickets with an incidence rate of one out of 25,000 babies. The mode of inheritance is X-linked dominant, i.e. with an affected father all daughters develop XLH and with an affected mother about 50 % of the children are affected. The most important biochemical feature of XLH is a marked and persisting hypophosphatemia which develops in babies at an age of 7 to 14 months. Prominent clinical symptoms are severe skeletal deformities, genu varum, short stature and teeth destructions. Patients with XLH require life long treatment. This usually consists of high dietary supplements of phosphate and massive doses of vitamin D or vitamin D metabolites.

The biochemical, radiological, and clinical symptoms of XLH-mice are very similar to those observed in humans with XLH. In experiments with XLH-mice using isolated brush border membrane vesicles from proximal tubules of the kidney the defect of the phosphate transport mechanism has been characterized. It was also found that the regulatory response of vitamin D-hormone (calcitriol; $1,25-(OH)_2D_3$) to various physiological signals (e.g. low dietary calcium and/or phosphate or parathyroid hormone) is altered. These findings have raised a debate as to whether the defective transport of phosphate in the proximal tubule is the primary pathogenetic factor of the disease or whether the defective tubular transport of phosphate is secondary in nature resulting from a defective setpoint control of hormones involved in calcium and phosphate metabolism. Experiments carried out with XLH-mice have significantly contributed to our understanding of the cellular processes and the subcellular mechanisms which are responsible for the development of the diseases symptoms. Results obtained from studies with the mouse model provide a rational basis for the development of a better and more effective treatment.

Pseudo-vitamin D-deficiency (PVDR) rickets is an inherited disorder of vitamin D metabolism which occurs in humans and pigs. The disease is inherited in both species by an autosomal recessive trait. Clinical and biochemical symptoms are also virtually identical in humans and pigs. Affected individuals develop symptoms of florid rickets at an age of four to eight months (humans) and in piglets at an age of one to eight weeks. In most cases florid rickets develops in piglets at an age of about four to five weeks. Affected human patients and piglets can effectively be treated by massive doses of vitamin D (about 100 to 1,000 times the physiological requirement of 12 to 15 µg at four weeks intervals).

In experiments with PVDR-piglets, the absence of the renal cortex 1α-hydroxylase has been identified as the etiological factor of the disorder. Studies with PVDR-piglets have also provided information why first clinical symptoms in piglets do not appear before an age of four to five weeks. It was found that plasma calcium and phosphate concentrations are normal in newborn homozygote PVDR-piglets, and that important mechanisms for maintenance of Ca and Pi homeostasis; e.g. transplacental transport of inorganic phosphates (P_i) and calcium(Ca), intestinal absorption of Ca and P_i; are at least partly independent from vitamin D during the first weeks of life. In experiments with the pig model an impaired metabolism of glucose was also identified. This disturbance was caused by a diminished glucose induced release of insulin and by a markedly reduced utilization of glucose by peripheral tissues. The findings obtained from experiments with PVDR-piglets provide a better understanding of many symptoms associated with PVDR and provide a basis for more effective therapeutic measures. In addition, animal models which suffer from distinct metabolic diseases are of great interest for the study of other basic problems which are often related to other human diseases.

There are distinct animal models for certain inborn human diseases

Experimental animals have played an important role in the past for elucidation of pathomechanisms of human metabolic bone diseases as well as for development of effective treatments. Many different forms of metabolic bone diseases exist in humans such as osteoporosis or various forms of vitamin D-resistant rickets. Many of these diseases are chronical, slow progressing processes. This feature makes it difficult to identify the causative factors which are responsible for the onset of such diseases. Metabolic bone diseases may be acquired or inherited. Table 1 lists a number of inherited diseases of

Table 1. Human inherited osteomalacic disorders involving vitamin D.

	animal model
1. Pseudo-vitamin D deficiency rickets, type I	pig
2. Pseudo-vitamin D deficiency rickets, type II	marmoset
3. X-linked hypophosphatemic rickets	mouse
4. Autosomal dominant hypophosphatemic rickets	-------
5. Hypophosphatasia	-------
6. Sarcoidosis	-------
7. Fanconi syndrome	-------

defective bone mineralization which occur in humans. One common feature of these diseases is a primary or secondary disturbance of vitamin D-metabolism, making the patient dependent on supraphysiological doses of vitamin D.

Most of the diseases listed in Table 1 are life threatening conditions. They usually require life long treatment. Considerable advances in prophylaxis and treatment of some of these disorders have been made in the past by the study of specific animal models.

Animals exhibiting diseased conditions very similar to those observed in humans, have been described for two of the disorders listed in Table 1, i.e. X-linked hypophosphatemic rickets and pseudo-vitamin D-deficiency rickets, type I. These diseases which were first described in humans also occur in mice and pigs, respectively. Most, if not all, of the clinical, pathological and biochemical symptoms described so far in the diseased animals resemble those found in human patients. In fact, most of the disease symptoms appear to be identical in the animals and humans.

This paper attempts to highlight the role these animal models have played to discover pathomechanisms and etiologies of the two disorders. A more comprehensive discussion of the topic will be presented elsewhere (HARMEYER 1989).

X-linked hypophosphatemic rickets in humans (XLH)

XLH is the most common form of inherited vitamin D-resistant rickets in humans (HERBERT 1986). The defect is present in one out of 25,000 babies (KRUSE 1985). Patients who suffer from the disease develop first clinical symptoms at an age of about 7 to 14 months (KRUSE & KUHLENCORDT 1980). Early symptoms are growth retardation and skeletal deformities (Fig. 1; BURNETT et al. 1964). The bone deformations usually start at the extremities but affect the whole skeleton when the disease progresses (Fig. 2; PEDERSEN & MCCARROLL 1951).

Fig. 1 Fig. 2

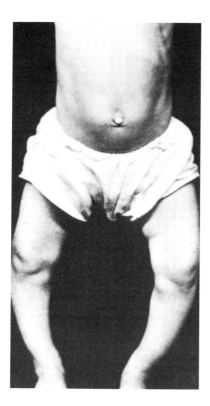
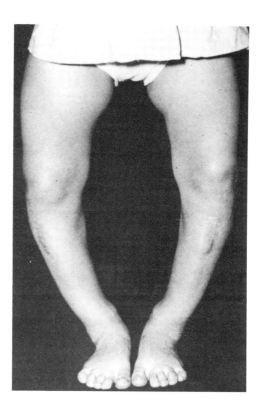

Fig. 1. *Two and a half years old girl with marked bowing of both legs and other signs of XLH (TAPIA et al. 1964).*

Fig. 2. *Thirteen years old girl with symptoms of XLH. The girl had already bilateral osteotomies of her tibias at an age of nine years. But leg deformities and genu varum had recurred since therapy with massive doses of vitamin D has been interrupted (TAPIA et al. 1964).*

The **clinical** symptoms are accompanied by **radiological** signs of rickets (Fig. 3). The mineral content of trabecular bone gradually diminishes while the cortical portions of the long bones develop a thicker matrix which is associated with reduced elongation (SCRIVER et al. 1982). Severe destructions of teeth also develop which show a typical histological feature, known as interglobular dentin (HERBERT 1986, TRACY et al. 1971).

The predominant **biochemical** feature of XLH is the persisting hypophosphatemia. In XLH-patients the plasma phosphate concentration declines from normal values of about 3.7 mmol/l to values of 2.0 mmol/l. Hypophosphatemia has proved to be remarkably refractory to different kinds of treatment. Clinical studies with XLH-patients have shown that hypophosphatemia is caused by a defect of renal reabsorption of phosphate. The renal transport capacity for phosphate is reduced to about 30 to 50 % of that of normal individuals.

The defective renal reabsorption of phosphate causes a permanent loss of the filtered phosphate load from plasma. The renal loss of the filtered load of phosphate increases from 5 to 20 % found in normal individuals to about 30 to 80 % in XLH-patients (BEAMER et al. 1979). An appropriate measure

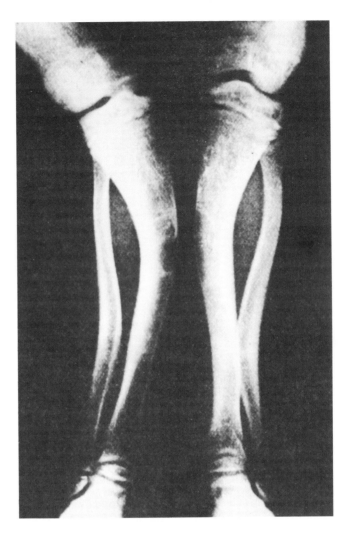

Fig. 3. Roentgenogram of the lower extremities of a six years old boy with XLH-symptoms after continuous treatment with 1.5×10^5 I.U. of vitamin D/d for about 3 years. The therapy has returned width of the epiphyseal lines and bone mineral content to almost normal. But the deformities and marked abnormalities in bone texture remained unaffected by the therapy (ALBRIGHT et al. 1937).

of the defect of renal excretion of phosphate in XLH- patients is the ratio of the maximally transported phosphate (TmP_i), mmoles per 100 ml of glomerular filtrate (GFR). The ratio, (TmP_i/GFR) has decreased in XLH-patients from about 1.27 to 1.87 mmoles/100 ml present in normal children to about 0.48 mmoles/100 ml (RASMUSSEN & ANAST 1983). Thus, elevation of the renal clearance of phosphate is the most prominent biochemical feature of XLH. It is this symptom which led FANCONI & GIRARDET (1952) to call the disease "Phosphate Diabetes". These authors were about the first who presented a more comprehensive description about the clinical and biochemical symptoms associated with the disease and who characterized the syndrome as a genuine disease. In retrospect, however, many cases of vitamin D-resistant rickets in children reported in the medical literature of the 40's and early 50's, yet unidentified at that time can now be addressed as XLH (ALBRIGHT et al. 1937).

Besides persisting hypophosphatemia from a pathomechanical point of view several other biochemical findings are of interest. The plasma calcium concentration for example is normal to slightly subnormal (SCRIVER et al. 1982). The parathyroid hormone concentration in plasma also appears to be unaltered (ARNAUD et al. 1971, CHAN et al. 1985). Moreover, the concentration of vitamin D-hormone (calcitriol) in plasma is also in the normal range (DREZNER et al.. 1980).

From a series of investigations carried out on human subjects with XLH a provisional pathogenetic concept has emerged which proved to be suitable to adequately explain the clinical and biochemical symptoms of the disease (RASMUSSEN & ANAST 1983, SCRIVER et al. 1982). According to this concept intestinal uptake of Ca and P_i is essentially unaltered (GLORIEUX et al. 1976). The principal factor which causes hypophosphatemia is the unphysiological high loss of renal phosphate. Hypophosphatemia in turn is considered to be responsible for the lowered accretion and the enhanced resorption rate of bone mineral and thus creating the skeletal lesions.

Other studies involving surveillance of affected pedigrees revealed an **X-linked dominant** mode of inheritance of the disease (RASMUSSEN & ANAST 1983). This is deduced from the observation, that with an affected father as a hemizygote, all daughters become heterozygotes and develop intermediate degrees of symptoms whereas all sons are normal. With an affected mother 50 % of the children (sons and daughters) develop symptoms of XLH (Fig. 4 and 5). However, the term "dominant" is not entirely appropriate, in that expression of symptoms in the homozygous and heterozygous female XLH genotype obviously differ. The homozygous (Hyp/Hyp) female is not existent. Only the Hyp/+ female survives. The difference in the expression of symptoms between the two female genotypes and between the heterozygote male and female shows, an intermediate type of inheritance of the symptoms.

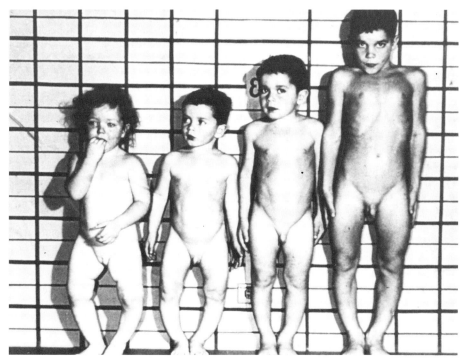

Fig. 4. XLH in three male brothers and a female first cousin. Marked leg deformities with genu varum, short stature and tibial torsions are visible (FRASER & SCRIVER 1979).

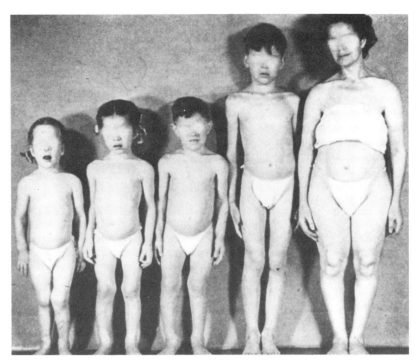

Fig. 5. A family with four out of eight children. The three boys with six, seven and eight years of age show clinical symptoms of XLH. The fourth child who is nine years old has no clinical symptoms of rickets. The mother had eight osteotomies for correction of her bowed legs when she was young. Rickets in the mother is now inactive (PEDERSEN & MCCARROLL 1951).

X-linked hypophosphatemic rickets in mice (Hyp/Y-mice)

EICHER et al.1976 described a disease in mice whose clinical, radiological and biochemical features were remarkably similar to those seen in humans with XLH. Three areas of research with XLH-mice will be briefly discussed here :
1) characterization of the animal model,
2) evaluation of the pathogenesis of symptoms,
3) therapeutic studies.

Characterization of the animal model

Continuing studies with this animal model showed that all the symptoms investigated, including the X-linked dominant mode of inheritance, appeared to be virtually identical to those found in human patients (TENENHOUSE et al. 1981, MEYER et al. 1979). The striking similarity included **clinical** and **radiological** features, such as dwarfism, kyphosis, a rachitic rosary and the lowered bone mineral content (Fig. 6). It also included similarity of **biochemical** symptoms, such as sustaining hypophosphatemia (with plasma phosphate concentrations around 1.1 mmol/l compared to 2.5 mmol/l of control mice) and the persisting renal loss of phosphate (TENENHOUSE et al. 1978).

Evaluation of the pathogenesis of symptoms

With the discovery of XLH in mice, it was immediately recognized that the animals might serve a model function for the study of problems associated with human XLH. In fact, since the Hyp/Y-mice

Fig. 6. Skeletal preparations from two male XLH mice (left) and one normal littermate, four months old. The Hyp/Y-male in the middle was offered a phosphate supplemented drinking water (6.75 g Na_2HPO_4 and 2.0 g KH_2PO_4/l) since he was one month old. Growth retardation and kyphosis is less pronounced than with the left Hyp/Y-animal which was maintained on a normal diet (EICHER et al. 1976).

have first been described intensive studies of pathogenesis and etiology of XLH have begun. Using isolated brush border membrane vesicles it was found for example that the defective transport of phosphate was confined to the PTH-sensitive part of the proximal tubule (GIASSON et al. 1977, TENENHOUSE & SCRIVER 1978, TENENHOUSE 1981). More importantly, the plasma calcitriol concentration - although being in a physiological range - was too low in view of the marked hypophosphatemia (DREZNER et al. 1980). In addition, stimulation of the renal 1α-hydroxylase through physiological signals, such as an injection of parathyroid hormone or through dietary P_i depletion, resulted in an increase in plasma calcitriol concentration too low compared to what was expected and what was seen with normal mice (MEYER er al. 1980). Conversely, the depressive effect of a high intake of calcium on the plasma 1,25-$(OH)_2D_3$ concentration was also too low compared to observations made with normal mice. These findings indicated a possible regulatory defect of vitamin D-metabolism in XLH. Further support for this assumption came from studies with XLH-mice which showed that stimulation of the renal 24-hydroxylase either by Ca or by 1,25-$(OH)_2D_3$ led to an unphysiologically high increase in the plasma 24,25-$(OH)_2D_3$ concentration (CUNNINGHAM et al. 1983). It was concluded from these findings and from the observation that the renal tubular reabsorption of phosphate in XLH-mice did not adequately respond to a lowering of the dietary intake of phosphate (MÜHLBAUER et al.. 1984), that the set point control of renal vitamin D-metabolism was defective in XLH (NESBITT et al. 1986, NESBITT et al. 1987, POSILLICO et al. 1985, TENENHOUSE & HENRY 1985, TENENHOUSE & JONES 1987). Other studies showed however, that vitamin D-metabolism in Hyp/Y-mice was unaltered under basal conditions (MEYER et al. 1984).

Based on this body of information, obtained from XLH-mice the classical hypothesis of a primary defect of the renal tubular phosphate transport system has been questioned (MEYER 1985). However, at the present ist is not clear which definite role the regulatory disturbance of vitamin D-metabolism plays in the developmental process of clinical and biochemical symptoms associated with XLH.

Other investigations with Hyp/Y-mice have provided evidence that the defective phosphate transport gene is expressed also in other tissues than in cells of the proximal tubule of the kidney (MEYER et al. 1984, TENENHOUSE et al. 1981).

Therapeutic studies

A remarkable feature of XLH patients is the considerable refractoriness to high doses of vitamin D (BURNETT et al. 1964, CHAN et al. 1985). Amounts of 2.5 to 25 mg/d, administered over months, have been tolerated without adverse effects on the systemic calcium homeostasis. Such high doses are often necessary to prevent development of rachitic lesions of bones (Fig. 7). The use of such high doses of vitamin D however, bears always the risk of hypercalcemia and renal damage (CURTIS et al. 1981). According to the current scheme of treatment which has been put forward by SCRIVER et al. (1982) vitamin D administration should be combined with high oral supplements of phosphate. Up to 5 g of elemental phosphate per day should be administered to human patients. The supplements should be given in several portions at about 4 hours intervals to provide a continuous systemic supply of phosphate and to avoid osmotic irritation of the intestines by the high phosphate load. It is easy to understand that this treatment applied during the whole life may incommode patients. In addition, the treatment does not always completely prevent growth retardation and development of short stature. Interesting new therapeutic approaches of XLH have come from testing certain synthetic vitamin D-metabolites such as 1α-OHD$_3$ (PEACOCK et al: 1977, SEINO et al: 1980a, SEINO et al. 1980b).

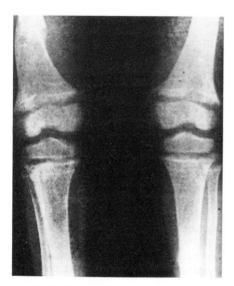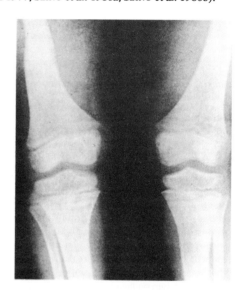

Fig. 7. (Left): about four years old boy with symptoms of XLH after bilateral osteotomy. The epiphyseal lines are widened and show active rickets. (Right): the same boy after four months treatment with about 1.5×10^5 I.U. (3.75mg) of vitamin D$_3$/d. Bone mineral content has decreased and the epiphyseal lines have reduced, but leg deformities are beginning to recur (PEDERSEN & MCCARROLL 1951).

There is however, no treatment available yet which converts the renal defect of phosphate transport in XLH-patients into normal. Nevertheless, Hyp/Y-mice have already played an important role to increase the understanding of pathomechanism involved in the development of symptoms with benefit for the human patient. The current studies with Hyp/Y-mice are primarily directed to further elucidate the hormonal, cellular, and subcellular nature of the observed defects. This experimental approach can possibly provide a more rational basis for the development of better and more effective treatments (MEYER 1985, TENENHOUSE et al. 1981).

Pseudo-vitamin D-deficiency rickets, type I (PVDR) in humans

PVDR is the second inherited disorder of the calcium and phosphorus homeostatic system which occurs in humans and in animals. Human PVDR is less frequent than human XLH. First **clinical**

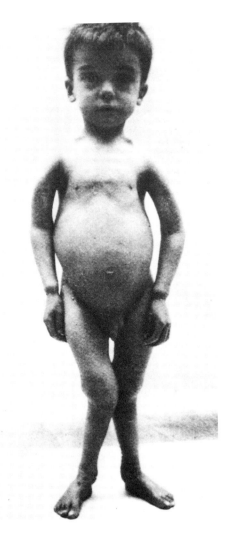

Fig. 8. Four years old child with PVDR. Note the compressed and severely deformed thorax, the marked deformities of legs and arms and the increased callus around the joints of legs and arms.

symptoms also develop in babies who carry the defective trait at an age of 4 to 8 months (KRUSE 1985). The symptoms are also growth retardation, (Fig. 8) bone deformities, (Fig. 9) severe swelling and pain of joints and bones, teeth destructions, (Fig. 10) progressing immobilization, fatigue, muscle weakness, and typical signs of florid rickets (PRADER et al. 1976, SCRIVER et al. 1982). **Radiological** signs of rickets or osteomalacia are a lower mineral content of bone, wider epiphyseal lines, a reduced calcification front, signs of osteopenia and a coarse trabecular structure. The latter symptoms result from the secondary hyperparathyroidism which in turn is induced by hypocalcemia. Affected human subjects usually die if they remain untreated. **Biochemical** findings are hypocalcemia, hypophosphatemia (SCRIVER 1970) and a marked increase in the activity of the alkaline phosphatase. These symptoms are typical for classical vitamin D-deficiency rickets. The concentrations of vitamin D and

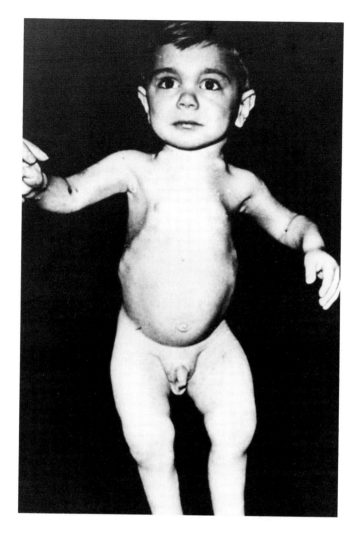

Fig. 9. Three years old child with PVDR showing severe deformities of long bones, "pot-bellied" abdomen, rachitic rosary and skeletal pain (FRASER & SCRIVER 1979).

of 25-OHD in plasma - with the latter being a specific indicator of the vitamin D-status - are normal however (SCRIVER et al. 1982). Vitamin D-intake with food and absorption of vitamin D from the intestine are also normal. Hence, the name "pseudo"-vitamin D-deficiency rickets appears appropriate. The calcitriol concentration in plasma is unphysiologically low (around 10 pg/ml compared to 50 to 70 pg/ml in control children). The parathyroid hormone concentration is elevated, associated with phosphaturia and hyperaminoaciduria. Mild plasma acidosis is seen sometimes. The calcemic response to Parathyroid hormone (PTH) injections is diminished. The intestinal absorption of calcium and phosphate is markedly reduced. For other symptoms see table 2.

The disease symptoms can completely be prevented by application of high doses of vitamin D (about 25 to 100 mg/(kg×d) or 25 OHD_3 (about ¼ of the vitamin D-dose) or by application of physiological amounts of 1,25-$(OH)_2D_3$ (FRASER & SCRIVER 1979, KRUSE & KUHLENCORDT 1980,

Table 2. Disease symptoms of PVDR in humans and piglets.

symptom	piglet	human
Clinical symptoms		
onset of first symptoms	3.5 to 5 weeks	3 to 6 months
skeletal deformities	extremities and backbone	extremities and backbone
retarded growth	present	present
Ca- and P-balance	diminished	diminished
irritability	present	present
mode of inheritance	autosomal-recessive	autosomal-recessive
Plasma		
calcium	diminished	diminished
phosphate	diminished	diminished
alkaline phosphatase	elevated	elevated
free amino acids	unchanged/diminished	not known
iPTH	elevated	elevated
vitamin D_3	normal	normal
25-OHD_3	elevated	normal
24,25-$(OH)_2D_3$	diminished	not known
25,26-$(OH)_2D_3$	normal	not known
1,25-$(OH)_2D_3$	diminished	diminished
Ca-response upon PTH-injection	not known	diminished
Intestine		
Ca- and P-absorption in newborn animals	normal	not known
Ca- and P-absorption during development of clinical symptoms	diminished	diminished
Ca-uptake (brush border membrane vesicles)	diminished	not known
ATP-dependent Ca-uptake (basolateral membrane vesicles)	normal	not known
amino acid-absorption	normal/diminished	not known
Na/P_i-cotransport (brush-border membrane vesicles, jejunum)	V_{max} diminished	not known
glucose absorption (brush border membrane vesicles, jejunum)	diminished	not known
nuclear 1,25-$(OH)_2D_3$-receptor	K_m and B_{max} same as in control piglets	not known
Ca-binding protein (calbindin)	diminished	not known

Table 2: cont.

symptom	piglet	human
Kidney		
P-excretion	elevated	elevated
TRP	diminished	diminished
amino acid excretion	generalized hyperaminoaciduria	generalized hyperaminoaciduria
cAMP-excretion	elevated	elevated
renal 25-OHD-1-hydroxylase	absent	not known
renal 25-OHD-24-hydroxylase	absent	not known
glomerular filtration rate	normal	not known
effective renal plasma flow	normal	not known
Na/P_i-cotransport (brush border membrane vesicles)	V_{max} diminished	not known
Bone		
Ca- and P-content	diminished	diminished
epiphyseal lines; Ca content of	diminished	diminished
teeth	enamel hypoplasia	enamel hypoplasia
Therapy		
physiological doses of vit.D_3 and 25-OHD$_3$	not effective	not effective
pharmacological doses of vit.D_3 and 25-OHD$_3$	effective	effective
physiological doses of 1,25-(OH)$_2$D$_3$	effective	effective
physiological doses of 1α-OHD$_3$	effective	effective
Other		
glucose induced response of insulin	diminished	not known
transplacental transport of Ca- and P	normal	not known
Ca- and P-homeostasis during pregnancy	partly independent from vitamin D	not known
peripheral utilization of glucose	diminished	not known

KRUSE 1985, SCRIVER et al. 1982). The findings indicate that the renal conversion of 25-OHD into 1,25-(OH)$_2$D is impaired.

The human disease was first characterized as an independent entity of heritable vitamin D-resistant rickets by PRADER et al.(1961). The genetic defect was shown to be transmitted to offspring by an **autosomal recessive** trait (RASMUSSEN & ANAST 1983). Clinical symptoms only develop in homo-

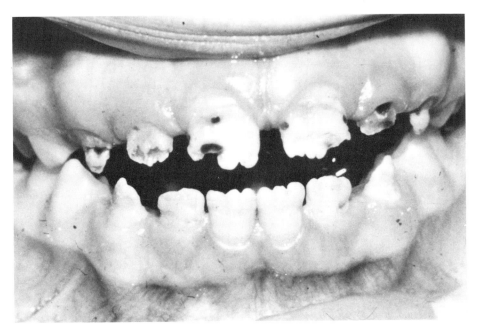

Fig. 10. Marked destructions of teeth occurring in a child with PVDR. The destructions comprise large pulp chambers, macroscopic discernible defects of the enamel structure (hypoplasia of enamel). (Courtesy of D. Fraser, Toronto).

zygote descendants. Heterozygotes are clinically and yet also biochemically indistinguishable from normal individuals.

Pseudo-vitamin D-deficiency rickets in piglets

A heritable rachitic condition with resistance to treatment with physiological doses of vitamin D was described in by PLONAIT (1962) in piglets of the German Landrace. The disorder was first denoted "vitamin D-resistant rickets of piglets". Breeding studies using affected animals revealed the autosomal recessive nature of the defect. Hundred percent of offspring from two affected parent animals was affected, but clinical symptoms did not develop in piglets generated by mating an affected animal (either sow or boar) with a normal control animal (MEYER & PLONAIT 1968). About 50 % of offspring which is generated by mating a heterozygote sow and a homozygote boar is affected with PVDR (Fig.11).

Piglets with PVDR die at an age of 10 to 12 weeks when they remain untreated. They can however, effectively be treated with intramuscular injections of high doses of vitamin D_3 (e.g. 10 to 20 mg per animal at 4 weeks intervals) (PLONAIT 1969). Vitamin D-treatment with such high doses is required for the whole life span. Various clinical, (Fig. 12 a and b) radiological and biochemical features of PVDR-piglets have been studied in more detail in recent years. To ease experimental work with this animal model the inherited trait was transmitted to minipigs (HARMEYER 1982b). It was recognized from the studies that the symptoms of PVDR-piglets resembled closely those observed in PVDR-human patients. In fact, all symptoms investigated so far are identical to those found in human PVDR (Table 2). Thus the animals constitute a true model for human PVDR.

Fig. 11. Mixed litter of three homozygote and two heterozygote piglets of German Landrace with PVDR. The animals are six weeks old. They were generated by mating a heterozygote sow with a homozygote boar.

Four areas of research for which PVDR-piglets have been used in the past will be discussed here:
1) treatment studies,
2) studies on vitamin D-metabolism,
3) investigation of pathomechanisms and etiology,
4) problems of vitamin D-function and bone metabolism.
Some relevant findings have emerged from these studies.

Treatment studies

Piglets with clinical symptoms of PVDR could effectively be treated with either pharmacological doses of vitamin D_3 (5 to 20 mg every 4 weeks) (PLONAIT 1963, PLONAIT 1965) or with pharmacolo-

Table 3. Plasma parameters of about six weeks old PVDR-piglets (n=15) and of age matched normal controls (n=10).

	PVDR	controls
vitamin D_3 (ng/ml)	5.1 ± 2.7	6.5 ± 2.3
25-OHD$_3$ (ng/ml)	38.5 ± 23.1	22.5 ± 18.4
24,25-(OH)$_2$D$_3$ (ng/ml)	4.5 ± 1.5	9.0 ± 3.1
25,26-(OH)$_2$D$_3$ (ng/ml)	4.5 ± 4.0	6.4 ± 4.5
1,25-(OH)$_2$D$_3$ (pg/ml)	26.1 ± 10.8	81.1 ± 35.9
calcium (mM)	1.8 ± 0.4	2.8 ± 0.3
inorganic phosphate (mM)	1.9 ± 0.6	3.0 ± 0.8
alkaline phosphatase (IU)	1335 ± 423	253 ± 90

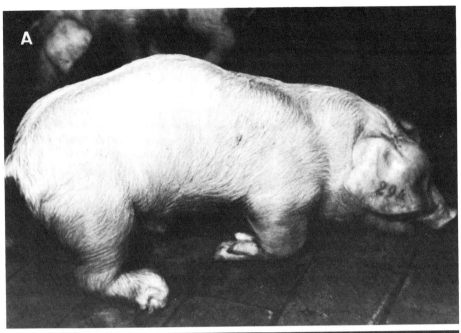

Fig. 12 A and B. Two seven weeks old homozygote untreated German Landrace piglets with PVDR. Both piglets show kyphosis, a pale skin and shaggy hair. The animal in Fig. 12a exhibits severe pain of joints and can no more stand on its legs. The animal will die within the next month when it remains untreated. The piglet in Fig 12b has been treated three days ago with 5×10^5 I.U. of vitamin D. Its physical condition has improved. But deformities of the backbone are largely irreversible.

gical doses of 25-OHD (3 to 10 mg every 4 weeks) (HARMEYER et al. 1977, HARMEYER 1982a). Effective treatment however, could also be achieved with injections of physiological amounts of either 1,25-$(OH)_2D_3$ (1 to 3 µg/d) or 1α-OHD_3 (3µg/d) (HARMEYER 1981). These findings indicated that renal calcitriol production was impaired and that clinical symptoms of rickets resulted from insufficient production of vitamin D-hormone. Administration of pharmacological amounts of vitamin D_3 or 25-OHD_3 healed not only the clinical symptoms of rickets but also reversed radiological and biochemical signs of rickets to normal. Treatment effects included correction of widened epiphysial lines, reduction of bone radiolucency, reversal of histological signs of bone mineralization (i.e. enlarged osteoid seams) and of mobilization of bone mineral (i.e. resorption lacunae). The lowered plasma Ca and P_i concentrations and the increased activity of alkaline phosphatase were also normalized (HARMEYER 1979) as were the elevated levels of circulating immunoreactive parathyroid hormone (DUCHATZ 1981, WILKE et al. 1979), the diminished intestinal absorption of calcium and phosphate (HARMEYER et al. 1976) the phosphaturia (HARMEYER et al. 1982) and the hyperaminoaciduria (HARMEYER & PLONAIT 1967). After adjustment of vitamin D doses to individual requirements the animals can be maintained for breeding with treatment at four weeks intervals. No clinical or biochemical signs of PVDR become apparent under such treatment.

Studies on vitamin D-metabolism

The concentrations of vitamin D_3 and of 25-OHD_3 in plasma of rachitic piglets were normal. The concentration of 25-OHD_3 was about 30 % above normal. However, the concentrations of both 1,25-$(OH)_2D_3$ and of 24,25-$(OH)_2D_3$ were unphysiologically low (Table 3) (KAUNE & HARMEYER 1986, KAUNE & HARMEYER 1987a). These findings indicated a partial block of the renal 1α-hydroxylase and 24-hydroxylase activity.

In vitro studies with isolated renal cortex homogenates and with mitochondrial renal cortex preparations showed however, a complete absence of both the 1α-hydroxylase and the 24-hydroxylase in such preparations (SCHREINER & HARMEYER 1986, WINKLER et al. 1982, WINKLER et al. 1986). No activity of the enzymes could be induced *in vitro* even by adding pharmacological amounts of substrate.

From treatment studies using massive doses of vitamin D_3 it was shown, however, that vitamin D administration led to a marked transient increase of both 1,25-$(OH)_2D_3$ and 24,25-$(OH)_2D_3$ in plasma. The observed concentrations of these metabolites exceeded temporarily their physiological levels (KAUNE & HARMEYER 1987b). We conclude from these observations that the responsive increase of the two metabolites as well as the low basal concentrations result from production by extrarenal sources (HARMEYER & KAUNE 1985). Although the complete absence of the two renal hydroxylases has not yet been demonstrated in human preparations from PVDR-patients the close resemblance of many other disease symptoms including the response to treatment with massive doses of vitamin D supports the assumption that a complete block of the renal enzymes is also present in humans.

Investigation of pathomechanisms and etiology

From the clinical studies carried out with human patients and from the results obtained with PVDR-piglets the following pathogenetic concept has been proposed (HARMEYER & KAUNE 1985). Lack of renal calcitriol production is the primary pathogenetic factor of PVDR. This defect creates metabolic deficiency of vitamin D-hormone and leads to reduced intestinal absorption of calcium and phosphate. This, in turn, creates hypocalcemia and hypophosphatemia. The diminished absorption of Ca and P_i lowers the concentrations of calcium and phosphate in plasma which in turn reduce the accretion rate of bone mineral in both the newly formed and the preformed bone matrix, inducing rickets and osteomalacia, respectively. Hypocalcemia also stimulates parathyroid hormone secretion and elevates the concentration of parathyroid hormone. Hyperparathyroidism induces increased resorption of bone mineral and elevation of the bone isoenzyme of alkaline phosphatase. It also stimulates cAMP mediated renal excretion of phosphate and amino acids. The concentration of cAMP in urine is elevated (HARMEYER & KAUNE 1985). Recent studies with PVDR-piglets have shown that hypocalcemia also significantly reduces utilization of glucose by peripheral tissues (SCHLUMBOHM et al. 1988). This observation provides a basis to explain at least partly the symptoms of fatigue, immobility and muscle weakness which are commonly observed in piglets and babies with symptoms of acute PVDR or D-deficiency rickets.

There is little information available yet about the allelic nature of the PVDR-defect. Marked variation has been observed in the time of first appearance of hypocalcemia in young piglets (between 6 to 40 days) and in the time of appearance of first clinical symptoms of rickets (between 1.5 to 8 weeks). This is true for animals which belong to the same litter and for individuals from different litters (LACHENMAIER-CURRLE 1985). The variation in the appearance of first clinical symptoms indicate a multiallelic nature of the genomic defect. But individual differences can also result from differences in food intake and the time of weaning.

Problems of vitamin D-function and of bone metabolism

Due to the absence of physiological concentrations of vitamin D-hormone, PVDR-pigs provide unique experimental conditions for the study of problems related to vitamin D-hormone functions and to the functions of systemic and cellular calcium. With the PVDR-piglet model it has been shown that vitamin D-hormone is not required for maintaining normal concentration gradients of calcium and phosphate across the placenta of pregnant sows (LACHENMAIER-CURRLE & HARMEYER 1988b). It has also been shown that vitamin D is not required for normal intestinal absorption of calcium and phosphate during the suckling period (LACHENMAIER-CURRLE & HARMEYER 1988a). Finally it was found that pregnant sows with PVDR are capable of maintaining normal calcium and phosphate concentrations in plasma when vitamin D-treatment is discontinued during pregnancy (LACHENMAIER-CURRLE 1985). Vitamin D-hormone and/or calcium, however, are required for normal glucose induced release of insulin by the β-cells of the endocrine pancreas (HARMEYER et al. 1985). An essential role of calcium for endogenous utilization of glucose has also been shown for the first time in studies with PVDR-piglets (SCHLUMBOHM & HARMEYER, in preparation).

General conclusions

Some inherited diseases in animals often resemble closely in their clinical, biochemical and pathogenetical features distinct familial human diseases. These inborn errors in animals serve as a model function for the corresponding diseases in humans and have played an important role in elucidating the pathogenetic mechanisms involved in the development of disease symptoms. Studies with animal models have contributed to identification of the etiology of the diseases in humans. The animal models have also been used for developing and testing suitable and effective treatments for the benefit of human patients. Moreover, such animal models sometimes also provide unique experimental conditions for the study of relevant basic physiological or clinical problems, others than those directly related to the diseased condition.

References

ALBRIGHT, F.; BUTLER, A.M.; BLOOMBERG, E. (1937): Rickets resistant to vitamin D therapy. Amer. J. Dis. Child. 54, 529 - 547.

ARNAUD, C.; GLORIEUX, F.; SCRIVER, C.R. (1971): Serum parathyroid hormone in X-linked hypophosphatemia. Science 173, 845 - 847.

BEAMER, W.G.; EICHER, E.M.; COWGILL, L.D. (1979): Familial hypophosphatemia (familial hypophosphatemic rickets). In: Spontaneous Animal Models of Human Disease. Vol. 2. (Editors: ANDREWS, E.J.; WARD, B.C.; ALTMAN, N.H.). Academic Press, London. Pp. 69-70.

BURNETT, C.H.; DENT, C.E.; HARPER, C.; WARLAND, B.J. (1964): Vitamin D-resistant rickets. Amer. J. Med. 36, 222 - 232.

CHAN, J.C.M.; ALON, U.; HIRSCHMAN, G.M. (1985): Renal hypophosphatemic rickets. J. Pediat. 106, 533 - 543.

CUNNINGHAM, J.; GOMES, H.; SEINO, Y.; CHASE, L.R. (1983): Abnormal 24-hydroxylation of 25-hydroxyvitamin D in the X-linked hypophosphatemic mouse. Endocrinology 112, 633 - 638.

CURTIS, J.; HSU, A.C.; BAUMAL, R.; RANCE, C.P.; STEELE, B.; KOOH, S.W.; FRASER, D. (1981): Risk of renal damage from large dose vitamin D therapy. Pediat. Res. 15, 506

DREZNER, M.K.; LYLES, K.W.; HAUSSLER, M.R.; HARRELSON, J.M. (1980): Evaluation of a role for 1,25-dihydroxyvitamin D3 in the pathogenesis and treatment of X-linked hypophosphatemic rickets and osteomalacia. J. Clin. Invest. 66, 1020 - 1032.

DUCHATZ, U. (1981): Veränderungen des Parathormonspiegels im Plasma bei Pseudo-Vitamin D-Mangelrachitis der Schweine vor und nach Therapie. Thesis, Tierärztliche Hochschule, Hannover.

EICHER, E.M.; SOUTHARD, J.H.; SCRIVER, C.R.; GLORIEUX, F.H. (1976): Hypophosphatemia: mouse model for human familial hypophosphatemic (vitamin D-resistant) rickets. Proc. Natl. Acad. Sci. USA 73, 4667 - 4671.

FANCONI, G.; GIRARDET, P. (1952): Familiärer persistierender Phosphatdiabetes mit D-vitamin-resistenter Rachitis. Helv. Paediat. Acta 7, 14 - 41.

FRASER, D.; SCRIVER. C.R. (1979): Disorders associated with hereditary or acquired abnormalities in vitamin D function: hereditary disorders associated with vitamin D resistance or defective phosphate metabolism. In: Endocrinology. Vol. 2 (Editors: DE GROOT, L. J.; C8AHILL, G.F., Jr; ODELL, W.D.; MARTINI, L.; POTTS, J.D., Jr.; NELSON, D.H.; STEINBERGER, E. ; WINEGRAD, A.J.). Grune & Stratton, N. Y.. Pp. 797- 807.

GIASSON , S.D.; BRUNETTI, M.G.; DANAN, G.; VIGNEAULT, N.; CARRIERE, S. (1977): Micropuncture study of renal phosphorus transport in hypophosphatemic vitamin D resistant rickets mice. Pflügers Arch. Eur. J. Physiol. 371, 33 - 38.

GLORIEUX, F.H.; MOREN, C.L.; TRAVERS, R.; DELVIN, E.E.; POIRIER, R. (1976): Intestinal phosphate transport in familial hypophosphatemic rickets. Pediat. Res. 10, 691 - 697.

HARMEYER, J. (1979): Charakterisierung eines erblichen Rachitismodells bei Schweinen. In: Wissenschaftlicher Arbeits- und Ergebnisbericht des Sonderforschungsberichs 146 "Versuchstierforschung", Tierärztliche Hochschule, Hannover. GFR. Pp. 59 - 77.

HARMEYER, J. (1981): Studies of vitamin D metabolism in vitamin D deficient pigs. In: Research Animals and Concepts of Applicability to Clinical Medicine, Symp. Sonderforschungsbereich 146 "Versuchsforschung", Tierärztliche Hochschule, Hannover.

HARMEYER, J. (1982a): Inherited disturbance of vitamin D metabolism in pigs. 2nd World Congr. Genet. Appl. Livestock Prod., Madrid, 2. Pp. 64 - 75.

HARMEYER, J. (1982b): Charakterisierung eines erblich bedingten Rachitismodells bei Schweinen. In: Wissenschaftlicher Arbeits-und Ergebnisbericht des Sonderforschungsbereichs 146 "Versuchstierforschung", Tierärztliche Hochschule, Hannover Pp. 78 - 94.

HARMEYER, J. (1989): Die Bedeutung von Tiermodellen bei der Erforschung menschlicher Krankheiten; drei Beispiele von Erkrankungen des Skeletts. In: Ergebnisse aus dem Sonderforschungsbereich "Versuchstierforschung" der Tierärztlichen und der Medizinischen Hochschule Hannover, Verlag VCH Weinheim, GFR.

HARMEYER, J.; GRABE, V. C.; MARTENS, H. (1977): Effects of metabolites and analogues of vitamin D3 in hereditary pseudo-vitamin Ddeficiency of pigs. In: Vitamin D, Biochemical, Chemical and Clinical Aspects Related to Calcium Metabolism. (Editors: NORMAN, A.W.; SCHAEFER, K.; COBURN, J.W.; DELUCA, H.F.; FRASER, D.; GRIGOLEIT, H.-G.; HERRATH, V.D.). Walter de Gruyter & Co., Berlin. Pp. 785 - 788 (Abstract).

HARMEYER, J.; GRABE, V. C.; WINKLER, I. (1982): Pseudo-vitamin D-deficiency rickets in pigs: an animal model for the study of familial vitamin D dependency. Exp. Biol. Med. 7, 117 - 125.

HARMEYER, J.; KAUNE, R. (1985): Vitamin D3-Metabolisierungsdefekt als Grundlage der erblichen Rachitis des Schweines. 16. Kongr. Dtsch. Vet. Med. Ges., Bad Nauheim, Parey, Berlin. Pp. 207 - 217.

HARMEYER, J.; KNORZ, S.; DWENGER, A.; WINKLER, I. (1985): The effect of vitamin D on the β-cell activity of the endocrine pancreas. Zbl. Vet. Med. A 32, 606 - 615.

HARMEYER, J.; PLONAIT, H. (1967): Generalisierte Hyperaminoacidurie mit erblicher Rachitis bei Schweinen. Helv. Paediat. Acta 22, 216 - 229.

HARMEYER, J.; VOGELGESANG, G.; GRABE, V.C. (1976): Measurement of intestinal calcium uptake in pigs with vitamin- D-dependent rickets. In: Nuclear Techniques in Animal Production and Health. Int. Atomic Energy Agency, Vienna. Pp. 243-252.

HERBERT, F.L. (1986): Hereditary hypophosphatemia rickets: an important awareness for dentists. ASDC J.Dent. Child. 53, 223 - 226.

KAUNE, R.; HARMEYER, J. (1986): Die quantitative Bestimmung von Vitamin D3 und seiner Metaboliten im Plasma. Biol. Chem. Hoppe-Seyler 367, 1135 - 1140.

KAUNE, R.; HARMEYER, J. (1987a): Eine erbliche Störung des Vitamin D Stoffwechsels beim Schwein, die Pseudo-Vitamin D-Mangelrachitis, Type I. Berl. Münch. Tierärztl. Wschr. 100, 6 - 13.

KAUNE, R. ; HARMEYER, J. (1987b): Vitamin D3 metabolism in a pig strain with pseudo-vitamin D-deficiency rickets, type I. Acta Endocrinol. 115, 345 - 352.

KRUSE, K. (1985): Hereditäre Störungen des Vitamin D-Stoffwechsels. Ergeb. d. inn. Med. und Kinderhkde. 54, 107 - 154.

KRUSE, H.P. ; KUHLENCORDT, F. (1980): Osteomalazie. In: Klinische Osteologie, Handbuch der inneren Medizin: Knochen, Gelenke, Muskeln, Teil 1B. 5. Auflage, Bd. 6. (Editors: KUHLENCORDT, F. & BARTELHEIMER, H.). Springer Verlag, Berlin. Pp. 751 - 820.

LACHENMAIER-CURRLE, U. (1985): Die Rolle von Vitamin D bei der Regulation des Calcium- und Phosphathaushaltes während der Perinatalperiode beim Schwein. Thesis, Universität Hohenheim, Fakultät Allg. und Angew. Naturwiss., GFR.

LACHENMAIER-CURRLE, U.; HARMEYER, J. (1988a): Intestinal absorption of calcium in newborn piglets, the role of vitamin D. Biol. Neonate 53, 327 - 335.

LACHENMAIER-CURRLE, U. ; HARMEYER, J. (1988b): Placental transport of calcium and phosphorus in pigs. J. Perinat. Med. (in press).

MEYER, H. ; PLONAIT, H. (1968): Über eine erbliche Kalziumstoffwechselstörung beim Schwein (erbliche Rachitis). Zbl. Vet. Med. A 15, 481 - 493.

MEYER, J.R.A.; JOWSEY, J.; MEYER, M.H. (1979): Osteomalacia and altered magnesium metabolism in the X-linked hypophosphatemic mouse. Calcif.Tissue Int. 27, 19 -26.

MEYER, R.A. (1985): Animal model of human disease: X-linked hypophosphatemia (familial or sex-linked vitamin-D-resistant rickets). Amer. J. Pathol. 118, 340 - 342.

MEYER, R.A.; GRAY, R.W.; MEYER, M.H. (1980): Abnormal vitamin D metabolism in the X-linked hypophosphatemic mouse. Endocrinology 107, 1577 - 1581.

MEYER, R.A.; MEYER, M.H.; GRAY, R.W. (1984): Metabolites of vitamin D in normal and X-linked hypophosphatemic mice. Calcif. Tissue Int. 36, 662 - 667.

MEYER, M.H.; MEYER, R.A.; IORIO, R.J. (1984): A role for the intestine in the bone disease of juvenile X-linked hypophosphatemic mice: malabsorption of calcium and reduced skeletal mineralization. Endocrinology 115, 1464 - 1470.

MÜHLBAUER, R.C.; BONJOUR, J.P.; FLEISCH, H. (1984): Abnormal hyperphosphatemic response to fasting in X-linked hypophosphatemic mice. Miner. Elektr. Metab. 10, 362 - 365.

NESBITT, T.; DREZNER, M.K.; LOBAUGH, B. (1986): Abnormal parathyroid hormone stimulation of 25-hydroxyvitamin D-1 alpha-hydroxylase activity in the hypophosphatemic mouse. Evidence for a generalized defect of vitamin D metabolism. J. Clin. Invest. 77, 181 - 187.

NESBITT, T.; LOBAUGH, B.; DREZNER, M.K. (1987): Calcitonin stimulation of renal 25-hydroxyvitamin D-1 alphahydroxylase activity in hypophosphatemic mice. Evidence that the regulation of calcitriol production is not universally abnormal in X-linked hypophosphatemia. J. Clin. Invest. 79, 15 - 19.

PEACOCK, M.; HEYBURN, P.J.; AARON, J.E. (1977): Vitamin D resistant hypophosphatemic osteomalacia. Clin. Endocrinol. 7, Suppl, 231S - 237S.

PEDERSEN, H.E. ; MCCARROLL, H.R. (1951): Vitamin-resistant rickets. J.Bone Joint Surg. 33-A, 203 - 220.

PLONAIT, H. (1962): Klinische Fragen der Calciumstoffwechselstörungen beim Schwein. Dtsch. Tierärztl. Wschr. 69, 198 - 202.

PLONAIT, H. (1963): Eine nicht ernährungsbedingte Rachitis beim Schwein. 17 Welt-Tierärzte-Kongress Hannover 1963. Pp. 1333 - 1334.

PLONAIT, H. (1965): Erbliche Rachitis bei Saugferkeln. Dtsch. Tierärztl. Wschr. 72, 255 - 256.

PLONAIT, H. (1969): Erbliche Rachitis der Saugferkel: Pathogenese und Therapie. Zbl. Vet. Med. A 16, 271 - 316.

POSILLICO, J.T.; LOBAUGH, B.; MUHLBAIER, L.H.; DREZNER, M.K. (1985): Abnormal parathyroid function in the X-linked hypophosphatemic mouse. Calcif. Tissue Int. 37, 418 - 422.

PRADER, A.; ILLIG, R.; HEIERLI, E. (1961): Eine besondere Form der priämren vitamin D-resistenten Rachitis mit Hypocalcämie und autosomal-dominantem Erbgang: die heriditäre Pseudo-Mangelrachitis. Helv. Paed. Acta 16, 452 - 468.

PRADER, A.; KIND, H.P.; DELUCA, H.F. (1976): Pseudo-Vitamin D deficiency (vitamin D dependency). In: Inborn Errors of Calcium and Bone Metabolism. (Editors: BICKEL, H. ; STERN, I.). University Park Press, Baltimore. Pp. 115-123.

RASMUSSEN, H. ; ANAST, C. (1983): Familial hypophosphatemic rickets and vitamin D-dependent rickets. In: The Metabolic Basis of Inherited Disease. (Editors: STANBURY, J.B.; WYNGAARDEN, J.B.; FREDRICKSEN, D.S.; GOLDSTEIN, J.L.; BROWN, M.S.). McGraw-Hill Book Comp, N.Y. Pp. 1743 - 1773.

SCHLUMBOHM,; HARMEYER, J.; C.; DWENGER, A: (1988): The effect of hypocalcemia and vitamin D-deficiency on glucose utilisation in piglets. In: Vitamin D. Molecular, Cellular and Clinical Endocrinology. (Editors: NORMAN, A.W.; SCHAEFER, K.; GRIGOLEIT, H.G.; HERRATH, v.D.). Walter de Gruyter & Co., Berlin. Pp. 889-890 (Abstract)

SCHREINER, F. ; HARMEYER, J. (1986): Charakterisierung der renalen 25-Hydroxycholecalciferol-1-Hydroxylase beim Schwein. J. Vet. Med. A 33, 746 - 756.

SCRIVER, C.R. (1970): Vitamin D dependency. Pediatrics 45, 361 - 363.

SCRIVER, C.R.; FRASER, D.; KOOH, S.W. (1982): Heriditary rickets. In: Calcium Disorders. (Editors: HEATH, D. ; MARX, S.J.). Butterworth Scientific, London. Pp. 1 - 46.

Seino, Y.; Shimotsuji, T.; Ishida, M.; Ishii, T.; Yamaoka, K.; Yabuuchi, H. (1980a): Vitamin D metabolism in hypophosphatemic vitamin D-resistant rickets. Contr. Nephrol. **2**, 101 - 106.

Seino, Y.; Shimotsuji, T.; Ishii, T.; Ishida, M.; Ikehara, C.; Yamaoka, K.; Yabuuchi, H.; Dokoh, S. (1980b): Treatment of hypo-phosphatemic vitamin D resistant rickets with massive doses of 1-α-hydroxyvitamin D_3 during childhood. Arch. Dis. Childh. **55**, 33 - 44.

Tapia, J.; Stearns, G.; Ponseti, I. (1964): Vitamin-D resistant rickets: a long-term clinical study of eleven patients. J. Bone Joint Surg. **46**, 935 - 958.

Tenenhouse, H.S.; Cole, D.E.C.; Scriver, C.R. (1981): Mendelian hypophosphataemias as probes of phosphate and sulphate transport by mammalian kidney. In: Inherited Problems of Transport, **17**. (Editors: Belton, N.; Toothill, C.). University Park Press, Baltimore. Pp. 231 - 262

Tenenhouse, H.S. ; Henry, H.L. (1985): Protein kinase activity and protein kinase inhibitor in mouse kidney; effect of the X-linked hyp mutation and vitamin D status. Endocrinology **117**, 1719 - 1726.

Tenenhouse, H.S. ; Jones, G. (1987): Effect of the x-linked hyp mutation and vitamin D status on induction of renal 25-hydroxyvitamin D_3-24-hydroxylase. Endocrinology **120**, 609 - 616.

Tenenhouse, H.S. ; Scriver C.R. (1978): The defect in transcellular transport of phosphate in the nephron is located in brush-border membranes in X-linked hypophosphatemia (Hyp mouse model). Can. J. Biochem. **56**, 640 - 646.

Tenenhouse, H.S.; Scriver, C.R.; MC Innes, R.R.; Glorieux, F.H.(1978): Renal handling of phosphate *in vivo* and *in vitro* by the X-linked hypophosphatemic male mouse: evidence for a defect in the brush border membrane. Kidney Int. **14**, 236 - 244.

Tracy, W.E.; Steen, J.C.; Steiner, J.E.; Buist, N.R.M. (1971): Analysis of dentine pathogenesis in vitamin D-resistant rickets. Oral Surg. Oral Med. Oral Pathol. **32**, 38 - 44.

Wilke, R.; Harmeyer, J.; Grabe, v.C.; Hehrmann, R.; Hesch, R.D.(1979): Regulatory hyperparathyroidism in a breed with vitamin D-dependency rickets. Acta Endocrinol. **92**, 295 - 308.

Winkler, I.; Grabe, v.C.; Harmeyer, J. (1982): Pseudo-vitamin D-deficiency rickets in pigs: *In vitro* measurement of 25-hydroxycholecalciferol-1-hydroxylase activity. Zbl. Vet. Med. A **29**, 81 - 88.

Winkler, I.; Schreiner, F.; Harmeyer, J. (1986): Absence of renal 25-hydroxycholecalciferol-1-hydroxylase activity in a pig strain with vitamin D-dependent rickets. Calcif. Tissue Int. **38**, 87 - 94.

Authors' address:
J. Harmeyer and R. Kaune, Physiologisches Institut der Tierärztlichen Hochschule Hannover, Bischofsholer Damm 15, D-3000 Hannover 1, F.R.G.

The pig and its plasma lipoprotein polymorphism in studies of atherosclerosis

J. RAPACZ and J. HASLER-RAPACZ

Introduction

Heart disease and stroke account for approximately half of all human deaths in the industrialized nations and atherosclerosis is the leading cause of these deaths. The atherosclerotic disease has reached an alarmingly high incidence, includes renal and peripheral vascular diseases, and is the major cause of morbidity and disability with aging. Atherosclerosis is visualized as a complex, multifactorial, and dynamic polypathogenic process involving vascular tissues affected by disturbances in metabolic processes leading to arterial lesions, which develop into a severe disease. Owing to these complexities the disease process is still poorly understood. Although in genetically susceptible animals atherosclerotic lesions can be induced and accelerated experimentally by a single factor such as a cholesterol rich diet, in humans they begin in childhood, spontaneously and undetected, and develop irregularly over a long life span into a severe atherosclerotic disease.

This insidious process represents a number of distinctly different clinical and morphopathologic patterns having various rates of progression, essentially without showing clinical effects, until the artery is nearly occluded. This impedes blood circulation and triggers a sudden appearance of symptoms in the form of a heart attack, sudden death, stroke, angina or claudication. In addition to the common difficulties in detecting the stage of the preclinical lesions, there are individual inherent differences and innumerable exogenous factors making investigations of the atherosclerotic process in humans difficult or infeasible. Therefore, reliable experimental animal models are desirable in an attempt to further advance and elucidate the nature of the atherosclerotic process in humans.

Experimental models provide opportunities to explore underlying mechanisms of the disease process and to identify both exogenous and inherent endogenous factors which can modify metabolic processes. In addition, models enable studies of the alleged atherogenicity of numerous variables and associated events under selected experimental conditions, and allow for comparisons with the clinically observed disease in humans. A variety of animal models have been developed during the past four decades in response to these needs, and the pig became a prominent model.

Studies using swine models have contributed to a better understanding of human atherosclerosis. It is beyond the limits of this article to review all of these reports. The reader is directed to a recent review, more extensive than this presentation, citing and reviewing in brief all available reports and their reviews published in English regarding the subject of atherosclerosis in swine (RAPACZ & HASLER-RAPACZ 1989). The first part of this chapter is intended to present a brief summary, cite selected studies which characterized swine spontaneous and experimentally induced atherosclerosis, its resemblance to the human disease, and identify atherogenic variables, as well as describe characteristic alterations in plasma constituents and in the morphology of the arterial wall. In addition, an attempt is made to consider individual variations among experimental swine, which were observed in a number of these studies, but rather seldom commented on their possible nature as being genetic.

The second part is intended to present a summary of advances made in studies of genetic aspects of lipoprotein diversity and its possible association with atherosclerosis. Immunogenetic studies, carried out during the last two decades in this laboratory, provide insights into genetic diversity in protein constituents of the lipoproteins known as apolipoproteins (Apo). A number of lipoprotein variants (allotypes, epitopes or apolipoprotein markers) have been identified and represent genetic polymorphisms located in the structure of the apolipoproteins. Some of the apolipoprotein mutants were found associated with the development of accelerated and severe atherosclerosis producing lesions which resemble very closely advanced atherosclerosis in humans.

Early studies on naturally occurring atherosclerosis in pigs

Although, around 1500 LEONARDO DA VINCI used swine in medical studies demonstrating the movement of the heart during the cardiac cycle, the pig has seldom been used in cardiovascular research until 1954 when GOTTLIEB and LALICH examined 1,775 pig aortas from animals 4 months to over 3 years old and reported on naturally occurring atherosclerosis in swine. At that time, domestic swine were represented by a large number of breeds worldwide (a rough estimate by the authors is over 200 breeds), and provided a vast array of genetic variation between and within breeds (GOTTLIEB & LALICH 1954). This variation offered useful large animal material for research in atherosclerosis. Similarities in size, physiology, cardiovascular structure and coronary artery distribution with humans as well as an omnivorous nature made swine a very desirable species to explore further its usefulness as a model (FRENCH & JENNINGS 1965, BUSTAD & MCCLELLAN 1966).

Following this finding, studies were undertaken by a number of investigators who focused their efforts primarily on three objectives:
1) characterization of atherosclerosis by determining grossly and histologically the prevalence, arterial distribution and severity of spontaneous atherosclerosis;
2) comparison of swine and human spontaneous atherosclerosis; and
3) exploration of the experimental induction and acceleration of the atherosclerotic process by manipulating dietary components.

At least seven independent investigations were carried out on spontaneous atherosclerosis. They included 2,269 pigs of both sexes in age from birth to 14 years of age, fed a basic low fat diet (fat < 5%) and represented over nine breeds and their crosses (JENNINGS et al. 1961, SKOLD & GETTY 1961, FRENCH et al. 1963, FRENCH & JENNINGS 1965, ROBERTS & STRAUS 1965, ROWSELL et al. 1965).

Results of studies during the first 15 years showed that swine spontaneous atherosclerosis has a remarkable resemblance to the early features of the human disease, and that the disease can be accelerated by feeding diets high in saturated fat and cholesterol. Consequently, the pig became a very attractive model to study atherosclerosis during the last two decades. Studies were continued to determine the effects of different endogenous and exogenous factors, mechanisms and time required for the development of advanced atherosclerosis; however, the principal variable investigated was the level and type of dietary fat and cholesterol. Together, these investigations covered the entire life span of swine providing an opportunity for studying all stages of atherosclerotic changes. Morphological descriptions of spontaneous atherosclerosis in these reports were based on gross, light and electron microscopy examinations (FRENCH et al. 1963, FRENCH & JENNINGS 1965). The investigations concentrated on the aorta and its main branches, with special attention to the coronary arteries.

In summary, research data on spontaneous atherosclerosis in swine indicated that extensive advanced changes with complicated atherosclerotic lesions such as ulceration, thrombosis and hemorrhages into the lesions, as well as elevations in plasma cholesterol and lipoproteins, identified in humans, have not been observed in swine. However, gross and microscopic examinations of the arteries in swine revealed many characteristics which showed remarkable resemblance to early features of the human disease. This inference was based on the following indications:
1) approximately 50% of swine examined showed similarly as in humans inherent susceptibility to atherosclerosis;
2) the same arteries (aorta, coronary and cerebral) were severely affected by atherosclerosis in swine as in humans;
3) swine atherosclerosis began early in life as fatty streaks or intimal thickenings;
4) regions of intimal thickening were common in the coronary arteries, and changes identified as early atherosclerosis occurred in these regions;
5) the changes showed similar distribution at points of branching as in humans, and the lesions increased progressively with proliferation of elastic tissues with age;
6) the lesions occurred when the animals consumed low fat diets;
7) the changes were located first in the thoracic segment of the aorta, then in the abdominal portions, and developed later in the coronary arteries;
8) smooth muscle cells appeared to proliferate and migrate into the intima where plaques developed;

9) plaques had collagen, elastic tissue, smooth-muscle cells, fibroblast-like cells, fat-filled cells, and exhibited fibrous tissue proliferation. Larger lesions with considerable amounts of extracellular fat had the appearance of atheromatous softening.

Dietary induced atherosclerosis in swine

The observations that spontaneous arterial lesions in swine resemble the human disease stimulated interest to investigate the effects of different dietary ingredients, especially lipids alleged as atherogenic, their levels of intake and interactions on the development and acceleration of experimental atherosclerosis. Additional variables which could be associated, or indicative of the disease process, were considered in some of these studies and included standard breeds *versus* miniature swine, age, sex and castration, as well as the effect of diets on changes in serum lipids (mainly cholesterol) and lipoproteins, and changes in arterial morphology and composition (BARNES et al. 1959, REISER et al. 1959, ROWSELL et al. 1960, DOWNIE et al. 1963, MORELAND et al. 1963, LUGINBÜHL et al. 1969, HILL et al. 1971, ST. CLAIR et al. 1971, KHAN et al. 1977, REITMAN et al. 1982, LEE 1986). The primary objectives were to search for indications that would elucidate the relationships between the type and intake of the diet on the development and process of the disease. Although dietary fats have always been regarded as hypercholesterogenic and atherogenic factors, and therefore they were intensely studied variables, the effect of dietary proteins and carbohydrates on experimental atherosclerosis has received some attention (BARNES et al. 1959, MORELAND et al. 1963, GREER et al. 1966, ST. CLAIR et al. 1971).

Summarizing the selected studies, it appears that manipulation of ingredients in the diet had effects in the majority of experiments on the development of atherosclerosis in swine. Diets enriched in saturated fats had accelerating effects, except for two studies, whereas carbohydrates and proteins showed no influence on the development of lesions (BRAGDON et al. 1957, CALLOWAY & POTTS 1962, MORELAND et al. 1963, GREER et al. 1966, ST. CLAIRE et al. 1971). The extent of the disease was influenced by the types and contents of dietary lipids (REISER et al. 1959, ROWSELL et al. 1960, MORELAND et al. 1963, GREER et al. 1966, HILL et al. 1971). The addition of cholesterol to the diet accelerated the atherosclerotic process up to eight-fold, increasing its rate and severity (REISER et al. 1959, ROWSELL et al. 1965, DOWNIE et al. 1963, MORELAND et al. 1963, GREER et al. 1966, HILL et al. 1971, KHAN et al. 1977). Equally important was the demonstration that the dietary-induced atherosclerosis in swine closely resembled the naturally occurring form in humans with regard to patterns, morphology and topography, however, with one essential difference. Unlike the spontaneous form the experimentally induced disease was associated with some, or marked serum cholesterol and lipoprotein elevations (REISER et al. 1959, ROWSELL et al. 1960, MORELAND et al. 1963, FRENCH & JENNINGS 1965, GREER et al. 1966, HILL et al. 1971, ST. CLAIR et al. 1971, MAHLEY et al. 1975, KHAN et al. 1977, GERRITY et al. 1979, REITMAN & MAHLEY 1979, REITMAN et al. 1982). This difference could be interpreted as follows: the elevated cholesterol is an unlikely initiator of the disease but its elevation is associated with the acceleration of atherosclerosis (MORELAND et al. 1963).

While sex showed no effect on atherosclerosis (ROWSELL et al. 1960, DOWNIE et al. 1963, GREER et al. 1966), castration increased the susceptibility (KHAN et al. 1977). The response to dietary manipulation seemed to undergo considerable changes with age. The young pigs appeared susceptible (REISER et al. 1959, ROWSELL et al. 1960, HILL et al. 1971), the older less responsive, and the aged swine quite resistant to the effect of atherogenic diets (LUGINBÜHL et al. 1969). The matter of the standard size *versus* miniature swine was considered and seemed to favor miniature swine which were proved suitable and very economical breeds for studies of atherosclerosis (MORELAND et al. 1963, FLORENTIN et al. 1968). However, the Yorkshire breed on this continent (ROWSELL et al. 1960) and Large White swine as its counterpart in Europe (STUNZI 1965) were considered as the most susceptible (ROBERTS & STRAUS 1965, STUNZI 1965), and the Landrace as more resistant than the miniature swine to the development of atherosclerosis (STUNZI 1965, HILL et al. 1971). Finally, results of these studies indicate that swine like other species, showed a varying degree of susceptibility to the effect of atherogenic diets. However, the possible genetic nature of these variations has not been studied or addressed except for a single comment (ROWSELL et al. 1960).

Hypercholesterolemia, morphologic and metabolic factors in plasma and arterial wall and accelerated atherosclerosis

Studies on dietary induced atherosclerosis in swine have provided important preliminary data showing that diets enriched in fats accelerate atherosclerosis. Subsequent studies were concerned with confirmation of the original finding and with use of pigs to investigate acceleration and severity of atherosclerosis. Other objectives were to find out if the advanced disease could be produced in a short period of time in young pigs of different strains, in swine with von Willebrand's disease, a disorder affecting platelet function, by diets containing different levels and forms of fats, diets fed different periods of time, or diets combined with other factors, such as arterial intimal injury by X-irradiation, or balloon catheterization (FLORENTIN et al. 1968, HILL et al. 1971, GOLDSMITH & JACOBI 1978, GERRITY et al. 1979, LEE et al. 1971, LEE et al. 1979, MAHLEY 1979a, POWNALL et al. 1980, REITMAN et al. 1982, THOMAS et al. 1983, GERRITY et al. 1985, KIM et al. 1985, GRIGGS et al. 1986, LEE 1986, SCOTT et al. 1986a). Findings of these investigations indicated, as it is universally recognized today (MAHLEY 1979b), that severe dietary induced serum hypercholesterolemia increases low density lipoprotein (LDL) levels and promotes the development of advanced atherosclerosis in the susceptible animals (HILL et al. 1971, THOMAS et al. 1983, GRIGGS et al. 1986, LEE 1986), which was characterized by marked intimal proliferation, cholesterol crystal deposition and necrosis (HILL et al. 1971, MAHLEY 1979b, REITMAN et al. 1982), and in the advanced lesions by lipid-rich calcified necrotic debris resembling those in humans consuming high-fat-cholesterol diets (THOMAS et al. 1983). While the ballooned catheterization and von Willebrand disease syndrome did not show synergistic effects with diet, the atherogenic diet combined with X-irradiation elicited the most advanced atherosclerosis accompanied by many myocardial infarcts and sudden death (LEE et al. 1971).

The high degree of variability in atherosclerotic lesions and plasma cholesterol levels observed among swine was in agreement with earlier investigations regarding the magnitude of response to the dietary fat and cholesterol. Such variations were of common occurrence in other species, including humans, and in pigs extended from high responders to nonresponders even when fed 3% to 4.2% cholesterol (FLORENTIN et al. 1968, MAHLEY 1979a). These findings supported previous observations and suggest that the susceptibility to atherosclerosis has a **genetic basis** and is independent of feeding the atherogenic diet in pigs, as well as in humans; however, in the susceptible animals the diet greatly accelerates the disease process.

The development of models for accelerated atherosclerosis in swine provided opportunities to study mechanisms of atherogenesis and accompanying events in a short time span, which occur during the entire period of life in man. In addition, morphologic and metabolic factors and mechanisms recognized to play a part in the process of development of atherosclerotic lesions have been studied extensively in swine (FLORENTIN et al. 1968, THOMAS et al. 1983, KIM et al. 1985, SCOTT et al. 1986a, SCOTT et al. 1986b).

Results of these studies are in agreement with human data and indicate that atherogenic lesions, in both young and old swine, occur in the lesion-prone area of the distal portion of the abdominal aorta and proximal part of the coronary arteries, where the intima is thickened, from birth onwards, by multiple layers resulting from accumulation of smooth muscle cells, monocyte-like cells, extracellular elements such as collagen, elastic tissue and glycosaminoglycans. Other studies (GERRITY 1981, GERRITY et al. 1979, GERRITY et al. 1985) added new findings to the mechanism of atherosclerotic lesion formation and identification of the lesion-predisposed areas. These lesion-prone aortic areas showed preferential adherence to the endothelium by blood monocytes, which subsequently migrate into the intima and become the major source of foam cells in the early lesions (GERRITY 1981). The foam cells with accumulated lipids migrate from the lesion into the lumen. This suggested that the monocytes, which phagocytize lipids, may play a part in lipid clearance and early atherosclerosis. Prolonged hypercholesterolemia may overload the lipid clearance system, triggering lipid accumulation, cell necrosis and subsequent smooth muscle cell involvement. In addition, this may lead to the progression of advanced lesion formation by the appearance of new components such as chemotactic factors (for review see LEE 1986).

Hereditary aspects of hyperlipidemia and response to atherogenic diets

Among the classic risk factors associated with the development of atherosclerosis is hypercholesterolemia, which is highly correlated with elevations of low density lipoprotein (LDL) concentrations. Unfortunately, a large proportion of the incidence of coronary heart disease is not explained by the classic risk factors. As indicated in this review, variations in the susceptibility to the development of atherosclerosis and hypercholesterolemia in response to the atherogenic diet were also common in swine; however, the nature of these variations has not been addressed. A number of studies, including those on swine (LEWIS & PAGE 1956, ROTHSCHILD & CHAPMAN 1976, POND et al. 1986), revealed that genetic factors exert effects on the cholesterol and LDL concentrations (BROWN & GOLDSTEIN 1986, HAMSTEN et al. 1986). Except for relatively rare cases of monogenic hypercholesterolemia in humans (BROWN & GOLDSTEIN 1986) and rabbits (WATANABEE 1980), resulting in both species from defects in LDL receptor activity, the mechanisms as well as the molecular basis for the hyperlipidemia remain unexplained. Genetically mediated LDL and cholesterol elevations may result from the action of a number of genes, diet-genotype interactions, or interactions between products of two or more genes or genotypes, which carry similar or related physiological functions in metabolic processes of lipoproteins or lipids. Polygenic abnormalities in lipid and LDL metabolism may be common, but are difficult to identify without gene specific markers. This is further complicated by more recent indications that all plasma lipoproteins are metabolically related, exposed to different lipid enzymes and lipid transfer proteins (SEGRETS & ALBERS 1986).

Biochemical studies during the last decade have shown that apolipoproteins and enzymes, active at lipid interfaces, play a central role in lipid transport systems and lipid metabolism. Recent studies on various aspects of plasma lipoproteins, including physicochemical characterization, structure, metabolism, cell biology and molecular biology have contributed considerably to a better understanding of lipoproteins (for review see SEGRETS & ALBERS 1986). Prior to that, little progress was made in biochemical analysis of polypeptide composition and identification of mutations in the main plasma lipoprotein, LDL, due to poor solubility of its principal protein, apolipoprotein B (Apo-B). Another contributing factor was insufficient interest in and support for exploring genetic aspects of lipoproteins.

Plasma lipoproteins in swine fed standard low or high fat and cholesterol diets

Lipoproteins of all the species studied are water soluble, heterogenous, and the most complex blood plasma macromolecules, composed of lipids (cholesterol, phospholipids and triglycerides) and specific proteins termed apolipoproteins. They are classified on the basis of their properties such as density, size, electrophoretic mobility, apolipoprotein content, gel chromatography or affinity chromatography separations. From published data on a number of lipoprotein characteristics, including concentration, distribution, composition, metabolism, structure, heterogeneity, dyslipoproteinemia and physiological function, it is evident that lipoproteins represent an important biological system in lipid transport and metabolism. Through these attributes lipoproteins are strongly implicated in atherogenesis. The essential part of the lipoprotein is the apolipoprotein, which maintains the lipoprotein structure, regulates its metabolism, transports and redistributes lipids among various tissues and acts as cofactor for the main lipid enzymes (for review see MAHLEY et al. 1984). There are at least 12 apolipoproteins identified in humans (ALAUPOVIC 1982).

Swine plasma lipoproteins have been studied in normal physiological conditions (JANADO & MARTIN 1968, JACKSON et al. 1973, MAHLEY & WEISGRABER 1974, KNIPPING et al. 1975) and under experimentally induced changes of the metabolism by hypercholesterolemia, induced by feeding high fat and cholesterol diets (HILL et al. 1971, MAHLEY et al. 1975, MAHLEY 1979a, REITMAN & MAHLEY 1979, POWNALL et al. 1980, REITMAN et al. 1982). Due to a large number of studies reporting on various characteristics of swine lipoproteins only selected aspects, such as resemblance to human lipoproteins and dietary induced changes will be reviewed here. Some other aspects which were the subject of our studies on lipoproteins will follow. The reader is referred to two recent reviews, one on swine lipoproteins exclusively (JACKSON 1983) and the other on swine and other animals (CHAPMAN 1980).

Early investigations on the comparison of several mammalian lipoproteins revealed that swine bear the closest resemblance to humans with regard to distribution and composition of plasma lipoproteins (HAVEL et al. 1955). Later studies showed that swine serum contains similarly to humans the three major classes: very low density (VLDL), low density (LDL) and high-density (HDL) lipoproteins, which can be separated by ultracentrifugation, electrophoresis or gel filtration (JANADO & MARTIN 1968, MAHLEY & WEISGRABER 1974). Plasma cholesterol levels in fasted pigs, maintained on standard swine diets, showed considerable variations from 60-230 mg/100 ml, and these values are lower than in humans (for review see RAPACZ & HASLER-RAPACZ 1989). Age, after 3 months post birth, and sex seemed not to influence cholesterol concentrations. Our studies showed no differences regarding sex until 12 months of age when cholesterol starts to decline gradually in males. The lifetime lowest cholesterol level is observed at birth. A sharp increase in LDL and cholesterol concentrations occurs during nursing; mean total cholesterol rises from 55 to 195 mg/100 ml in plasma of normocholesterolemic pigs, and from 110 to 425 mg/100 ml in pigs with inherited hyper-lipoproteminia and cholesterolemia (RAPACZ & HASLER-RAPACZ 1984, RAPACZ et al. in preparation). A rapid decrease takes place post weaning and is followed by a gradual recovery reaching a steady state at 4 months of age. During the adult life females showed some variations in the LDL and cholesterol concentrations and this seemed to be correlated with the reproductive cycle. An extensive comparison of human lipoproteins with swine by two methods (electrophoresis and standard ultracentrifugation) showed that both lipoproteins are essentially identical with respect to chemical composition, immunochemical reactivity, size by electron microscopy, and apoprotein content by polyacrylamide gel electrophoresis (MAHLEY & WEISGRABER 1974). The differences between swine and human lipoprotein were:

1) extended LDL density (d) distribution in swine to d 1.09 g/ml (JANADO & MARTIN 1968);
2) absence or very low concentration of apolipoprotein-AII (JACKSON et al. 1973); and
3) prominence of apo-CII in VLDL (KNIPPING et al. 1975).

The presence of two LDL's in swine, LDL1 and LDL2, was confirmed by several studies (MAHLEY & WEISGRABER 1974, KNIPPING et al. 1975, CHAPMAN 1980, JACKSON 1983).

Feeding atherogenic diets, high in fat and cholesterol, led in the susceptible animals to moderate or severe hypercholesterolemia and accelerated atherosclerosis. The hypercholesteremia was always correlated with LDL elevations and caused dramatic changes in the distribution and size of LDL1 and LDL2 (ROWSELL et al. 1960, HILL et al. 1971, ST. CLAIR et al. 1971, MAHLEY et al. 1975, GOLDSMITH & JACOBI 1978, GERRITY et al. 1979, MAHLEY 1979b, REITMAN & MAHLEY 1979, POWNALL et al. 1980, REITMAN et al. 1982, THOMAS et al. 1983, KIM et al. 1985, GRIGGS et al. 1986, SCOTT et al. 1986a). Extensive studies on miniature swine performed by MAHLEY and coinvestigators contributed to the finding that feeding high fat and cholesterol induces marked changes in plasma lipoproteins which are similar to both qualitative and quantitative changes in various species including humans (MAHLEY et al. 1975, MAHLEY 1979b, REITMAN & MAHLEY 1979, CHAPMAN 1980, POWNALL et al. 1980, REITMAN et al. 1982). These changes include:

1) appearance of β-VLDL in d < 1.006 g/ml fraction and a lipoprotein referred to as HDL$_c$ (cholesterol induced);
2) increase of LDL and intermediate density lipoprotein (IDL, d 1.006-1.02 g/ml); and
3) decrease in typical HDL (MAHLEY et al. 1975, MAHLEY 1979, REITMAN & MAHLEY 1979).

The β-VLDL liporotein is α-migrating, cholesterol rich VLDL, and the HDL$_c$ has α-mobility of d 1.02-1.087 g/ml, and while lacking Apo-B it is a cholesterol rich lipoprotein containing Apo-E, Apo-AI and the fast migrating C apoproteins (MAHLEY et al. 1975). These two lipoproteins, β-VLD and HDL$_c$, transport primarily dietary induced plasma cholesterol and were termed atherogenic lipoproteins. In our recently developed strain of swine with inherited hyper-LDL-cholesterolemia (IHLC) (RAPACZ & HASLER-RAPACZ 1984), which will be discussed later, elevated cholesterol is associated primarily with the buoyant fraction (CHECOVICH et al. 1988), or layers 2 and 3 of the gradient ultracentrifuge fractions (LEE et al. 1987a, LEE et al. 1987b).

Immunogenetically identified lipoprotein and DNA polymorphisms in swine

Search for genetic polymorphism of swine LDL was initiated following the discovery in humans of 10 immunological variants (epitopes, allotypes) designated Ag (BÜTLER et al. 1972). The Ag allotypes represent plasma LDL polymorphisms detected by the use of precipitating antibodies, evoked unintentionally in anemic patients receiving frequent blood transfusions containing immunologically incompatible proteins. Furthermore, anticipation of the importance of LDL in lipid transport and metabolism stimulated our interest in exploration of swine lipoprotein polymorphisms. Using the intraspecies immunization model (alloimmunization) for the development of epitope-specific antibodies, as demonstrated for Ag-LDL allotypes in humans, pigs were inoculated with serum LDL derived from swine of different breed origin. The assumption was that if mutations changed the amino acid sequence of Apo-B, the modified part of the protein, if antigenic, could elicit an antibody for the mutant epitope in pigs which lack this allotype.

Allospecific reagents distinguishing 22 LDL variants by gel precipitation reaction, designated Lpb (L-lipoprotein, p-pig, b-Apo-B), revealed that the Lpb epitopes represent differences in the structure of apolipoprotein-B (Apo-B, the predominant protein component of LDL (RAPACZ et al. 1970, RAPACZ et al. 1976, RAPACZ et al. 1978, RAPACZ 1982, RAPACZ & HASLER-RAPACZ 1984, RAPACZ et al. 1986b).

Genetic studies on over 6,000 pigs, representing 37 breeds, have shown that the allotypes behave as codominant traits and form eight distinct groups or phenogroups (haplotypes), transmitted from generation to generation as conserved units (Fig. 1), each determined by the corresponding *apo-B* allelic gene. Each of the eight haplotypes consists of one mutant (Lpb1-Lpb8) and seven common or shared epitopes (Lpb11-Lpb18), the latter presenting a non-mutant evolutionarily conserved part of Apo-B, which is immunologically related to human and other non-human primate Apo-B, whereas the mutant allotypes are species-specific and immunologically unrelated to Apo-B of other species (RAPACZ et al. 1976, RAPACZ et al. 1978, RAPACZ 1978, RAPACZ 1982, RAPACZ & HASLER-RAPACZ 1984).

No exception has been encountered with regard to the number and epitope specificities in the Lpb phenogroups (haplotypes) among almost 15,000 swine surveyed. Each of the 16 anti-Lpb reagents has its unique value in studying the genetic complexity of Apo-B, as well as in identifying new Lpb mutants (RAPACZ 1978, RAPACZ et al. 1978). Except for two very common alleles Lpb^5 and Lpb^8, others are found at low frequencies and their distribution is associated with specific breeds (RAPACZ 1974). Since no exception was found with regard to the number and specificity of epitopes in each set, we proposed that the original progenitor *apo-B* gene in swine contained genetic information only for the common allotypes, Lpb11-18. The individual epitopes appeared later during speciation, when single mutations at various parts of the *apo-B* gene took place, leading to the appearance of the code for the individual allotype and the loss of information for the corresponding common epitope. This process resulted in the formation of mutually exclusive pairs of epitopes; e.g., Lpb1 is the substitute for Lpb11, or Lpb7 for Lpb17 in the respective haplotype, and there are eight pairs of mutually exclusive allotypes identified. Although no comparable Apo-B epitope complexity to swine was found in other species, rhesus monkeys and chicken closely resemble the swine model (HASLER-RAPACZ & RAPACZ 1982, RAPACZ 1982). The genetic control of 10 human LDL markers designated Ag allotypes was proposed to be controlled by five pairs of allelic genes placed at closely linked loci (BÜTLER et al. 1972, RAPACZ 1978). Recent molecular genetic studies (BERG et al. 1986, MA et al. 1987) showed, as suggested earlier (RAPACZ et al. 1978, RAPACZ 1982), that the Ag system, similarly to Lpb in swine and Lmb in rhesus monkeys (HASLER-RAPACZ & RAPACZ 1982) is controlled by one complex locus, *apo-B*. It is likely that in each of the eight described Lpb haplotypes (Fig. 1) there are additional Lpb epitopes temporarily named semi-common allotypes (each occurring in at least two Lpb haplotypes) and are designated Lpb21-27 and Lpb31-37. These 14 new allotypes form seven pairs of mutually exclusive epitopes; e.g., Lpb21 with Lpb31 (RAPACZ 1978), arranged in eight sets. Each set carries seven semi-common epitopes and is linked to one of the eight Lpb haplotypes giving rise to 15 Lpb epitopes in each Apo-B haplotype.

Serological patterns of Lpb allotypes (Apo-B epitopes and the immunogenetic model of Lpb (Apo-B).

A   **B**

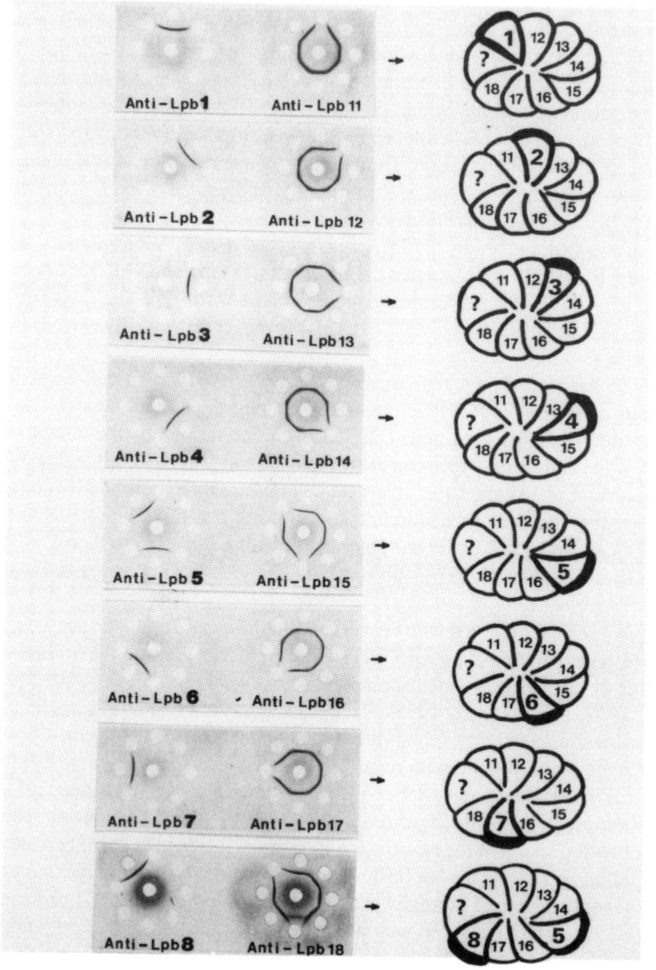

Fig. 1. Immunogenetic model for swine LDL polymorphism. (A) Patterns of precipitation reaction in agar gel of eight plasma samples (peripheral wells), representing eight Lpb homozygous genotypes, $Lpb^{1/1}$-$Lpb^{8/8}$ placed clockwise starting at the top of each rosette, respectively, reacted with 16 alloimmune reagents, anti-Lpb1-Lpb8 and anti-Lpb11-Lpb18, as indicated. Each reagent detects a different epitope in the Apo-B structure. Each of eight epitopes, designated as individual (mutant) allotypes, Lpb1-Lpb8, forms a mutually exclusive pair of epitopes with one of eight (Lpb11-18) corresponding common (ancestral) allotypes (e.g., Lpb2-Lpb12, or Lpb5-Lpb15). The Lpb phenogroup (haplotype) of each plasma is defined by the sum of the precipitation reactions with the Lpb reagents; e.g., the plasma in the top wells according to its reactions with the 16 reagents exhibits the Lpb1, 12, 13, 14, 15, 16, 17, 18 haplotype. The gel slides with the developed reactions were dried and stained for lipid with Sudan Black B. (B) Schematic representation of eight identified Lpb haplotypes (phenogroups), deduced from segregation data, each composed of Apo-B epitopes specified by the corresponding Lpb (apo-B) allelic gene designated by numerals, 1-8. Each allele of apo-B determines a complex haplotype bearing a set of at least eight distinct Apo-B epitopes.

Since each pig carries two *apo-B* alleles, one inherited from the mother and the other from theanimals (CHAPMAN 1980). father, the eight Lpb alleles $(Lpb^1\text{-}Lpb^8)$ yield a total of 36 *apo-B* combinations (genotypes) and 36 immunologically distinguishable phenotypes, each composed of two homomeric LDL molecules (RAPACZ et al. 1989a). Recent molecular genetic studies revealed that immunologically detectable Lpb epitopes are associated with restriction fragment lenght polymorphisms (RFLP's) of swine *apo-B* DNA. The sequenced part of the swine *apo-B* gene and the decoded amino acid sequence for that part showed 70% identity with humans (MAEDA et al. 1988).

In addition to the Lpb epitopes, six other lipoprotein allotypes have been reported by this laboratory, and include: Lps1, Lpt1, Lpr1, Lpr2, Lpu1 and Lpu2. Each allotype is determined by the individual gene; two autosomal dominant Lps^1 and Lpt^1, and two pairs of autosomal codominant alleles, Lpr^1, Lpr^2 and Lpu^1, Lpu^2, respectively. Homology of Lps1, Lpt1, Lpu1 and Lpu2 allotypes with any of the human apolipoproteins (ALAUPOVIC 1982) is not known, exept that the Lpr lipoprotein is immunologically related to human Apo-D (MCCONATHY & ALAUPOVIC 1973). Characterization of Lpr revealed that Apo-R forms an allelic multimer, which is distributed mainly in very high density lipoproteins (VHDL). The remaining Lpr occurs in VLDL, in complexes with β-VLDL (RAPACZ et al. 1986a). Plasma concentrations of Lpr1 in the $Lpr^{1/1}$ pigs is three times higher than Lpr2 in $Lpr^{2/2}$ animals. The Lpr^1 allele was found so far at a very low frequency and only in Chester White pigs. All swine allotypes identified (RAPACZ 1982) are associated with the LDL class, except Lpr1 and Lpr2 (RAPACZ et al. 1986a). The genetic loci for *Lpt* and *Lpu* are very closely linked with the *apo-B* locus. The Lpu^1 gene was found in a very small number of pigs, also of Chester White origin, and is linked with Lpb^5, forming the Lpb^5-Lpu^1 haplotype. The Lpt1 allotype may be associated with lipoprotein(a) (FLESS et al. 1984), may represent a second Apo-B like apolipoprotein (RAPACZ 1978), or an apolipoprotein not yet identified. The Lpt^1 occurs in linkage with all Lpb haplotypes, except Lpb^3.

Lpb polymorphism and the response to feeding high fat or atherogenic diet

Indications implicating Lpb alleles as genetic factors associated with either enhancing susceptibility or resistance to diet-induced cholesterolemia and/or atherosclerosis were obtained from two studies. Twelve of 22 Thompson miniature swine fed an atherogenic diet (25% fat, 2% cholesterol) for 7 months showed resistance to plasma cholesterol elevation while the remaining 10 swine were responders. The subsequent Lpb test revealed that all the resistant pigs were homozygous $Lpb^{3/3}$, whereas the responders were genotyped as heterozygous $Lpb^{3/5}$ (RAPACZ 1978). In the second study (RAPACZ et al. 1977) the effects of consuming a diet for 10 months, containing 19% fat with varying mixtures of *trans*-unsaturated, *cis*-unsaturated and saturated fatty acids were studied to determine whether the consumption of *trans*-unsaturated fatty acids by swine would induce more aortic lipidosis than saturated fatty acids. Aortas were evaluated from 57 pigs which were crossbreds carrying components of four breeds, and their plasma tested for Lpb genotypes and lipid levels. Statistical analysis of the data shows an effect of diet-Lpb genotype interaction ($p < 0.003$) on fatty streaking: pigs carrying the Lpb^5 allele had a greater tendency to develop fatty streaking than those with Lpb^8. The mechanisms underlying the associations in both studies are unknown.

Lipoprotein polymorphism and inherited hyper-LDL-cholesterolemia (IHLC) and accelerated atherosclerosis

The first marked elevation in LDL was observed in sera from several related Chester White pigs, all $Lpb^{5/5}$ or $Lpb^{5/8}$. Subsequently, tested serum samples for LDL and cholesterol concentrations from these pigs and their offspring confirmed the original observation. The results suggested that hypercholesterolemia and LDL elevations are genetically controlled. To test this hypothesis, a group of 12 swine served as original breeders in our Immunogenetic Project Herd (IPH) to produce five generations with a total of 175 closely related pigs (RAPACZ & HASLER-RAPACZ 1984). The cholesterol level increased through the generations reaching a mean of 266 mg/100 ml in the fifth generation. The mean cholesterol value and its standard deviation (SD) for all five generations was 176.5 ± 63.5

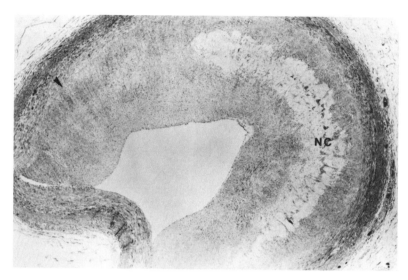

Fig. 2. *Coronary artery from a 29 month old hypercholesterolemic pig. There is an extensive intimal lesion above the internal elastic lamina (arrow) which consists of a lipid-rich necrotic core (NC), smooth muscle cells, macrophages and connective tissue. (Hematoxylin and eosin, 35x). Courtesy of Dr. Margaret F. Prescott and Christine H. McBride, Atherosclerosis Research, Ciba-Geigy Corp., Summit, NJ.*

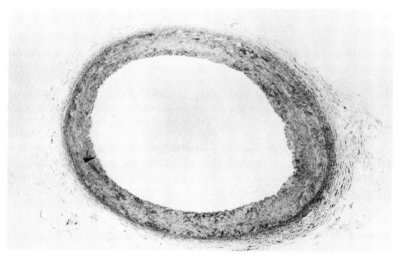

Fig. 3. *Coronary artery from a 30 month old normocholesterolemic pig. The intima consists of one layer of endothelial cells above the internal elastic lamina (arrow). (Hematoxylin and eosin, 35x). Courtesy of Dr. Margaret F. Prescott and Christine H. McBride, Atherosclerosis Research, Ciba-Geigy Corp., Summit, NJ.*

mg/100 ml. During 6 years of these studies we observed that the animals which had the highest LDL and cholesterol levels were invariably homozygous $Lpb^{5/5}\ Lpu^{1/1}$, and in addition they were heterozygous or homozygous for Lpr^1. Three additional generations were produced subsequently and the mean cholesterol value and its SD for a total of eight generations of $Lpb^{5/5}$, $Lpu^{1/1}$, $Lpr^{1/1}$ animals was 257 ± 67 mg/100 ml.

The final conclusion from these studies was that a genetically mediated spontaneous plasma low density hyperlipoproteinemia and hypercholesterolemia has been identified in a strain of swine developed in the experimental herd of the Immunogenetic Laboratory of the University of Wisconsin, Madison. The hyperlipoproteinemia phenotype is highly correlated with hypercholesterolemia and is associated with three lipoprotein allotypes, Lpb5, Lpu1 and Lpr1 (RAPACZ & HASLER-RAPACZ 1984). The cause of this elevation, or mechanisms underlying the associations are unknown. Segregation data from the progeny test indicated that the inherited hyperlipoproteinemia and hypercholesterolemia (IHLC) have a polygenic basis. Whether the Lpu1 and Lpr1 mutant epitopes contribute to the hypercholesterolemia in the Lpb^5 pigs remains to be determined. The mean plasma cholesterol value for pigs of very common $Lpb^{5/5}$, $Lpu^{2/2}$, $Lpr^{2/2}$ genotype, which were derived from six different breeds, was 105.4 ± 12.8 mg/100 ml. The mean cholesterol value for pigs carrying apo-B alleles other than Lpb^5 or Lpb^8 in the Lpb genotype, was 81.3 ± 11.6 mg/100 ml (RAPACZ & HASLER-RAPACZ 1984, RAPACZ et al. 1986b). The difference between the means for these two groups of Lpb genotypes is statistically significant. These results are interpreted as an indication that an inherent association exists between some homozygous Lpb genotypes, especially $Lpb^{5/5}$, and the distinct cholesterol and LDL concentrations. This contention is further supported by a recent observation on cholesterol and LDL concentrations in another Lpb homozygote, $Lpb^{7/7}$ (RAPACZ et al. 1986c). Due to a low gene frequency of Lpb7 only 24 swine of the $Lpb^{7/7}$ genotype were evaluated, however they all had the lowest cholesterol values, mean 69.8 ± 7.3 mg/100 ml (RAPACZ & HASLER-RAPACZ, unpublished). Although the cholesterol and LDL concentrations for the remaining Lpb homozygotes are not yet established, due to their low number, the observations on the relationship between the tested Lpb genotypes and LDL and cholesterol concentrations seem to indicate that the Apo-B polymorphism reflects on lipoprotein and cholesterol metabolism in swine. This finding is unique and of biological importance for future studies on lipoproteins, lipids and cholesterol metabolism and atherosclerosis, as well as for elucidations of apolipoprotein mutations in relation to physiological functions of lipoproteins.

High rates of sudden death during the development of the IHLC strain of swine, and a short life span (up to 4 years) prompted us to collect hearts for histological analysis to examine the arterial lesion involvement in normolipemic and hypercholesterolemic pigs (RAPACZ et al. 1986b, ROSENFELD et al. 1988) ranging in age from birth to 62 months. Pigs carrying the Lpb5 mutant had extensive fatty streaks and foam cell lesions at 7 months of age, and those between 7 and 14 months of age developed extensive lesions in all three coronary arteries. Animals between 21 and 32 months of age developed complex and large lesions restricting blood flow in the affected swine as shown in Fig. 2, while the coronary artery from a 30 month-old control pig exhibited practically no pathological involvement as shown in Fig. 3, a normal finding (GOTTLIEB & LALICH 1954, FRENCH & JENNINGS 1965, ROWSELL et al. 1965, RATCLIFFE et al. 1970). To better characterize atherosclerotic lesions in IHLC swine, immuno-cytochemical, histologic and morphologic analyses were conducted on arteries of the two oldest IHLC females, 48 and 54 months of age. Complex lesions in the coronary arteries contained foci of the following: inflammatory cells, smooth muscle cells with extensive connective tissue deposition, macrophage-derived foam cells, necrosis with extracellular lipid and calcification, neo-vascularization and hemorrhage (PRESCOTT et al., unpublished). Although numerous studies showed close resemblance between swine lesions and human early lesions, new features observed in this study are indicative that the spontaneous advanced atherosclerosis, not observed previously in swine, occurs in pigs with IHLC and resembles the human advanced disease. Unlike in familial hypercholesterolemia in humans (BROWN & GOLDSTEIN 1986) and rabbits (WATANABEE 1980) the IHLC pigs have normal LDL receptor activity, thus, this strain highlights the discovery of another genetic defect in lipoprotein and lipid metabolism. In all studies on dietary induced hypercholesterolemia and accelerated atherosclerosis in swine cited earlier, the hypercholesterolemia was of exogenous origin when the normal mechanisms of lipoprotein clearance were overwhelmed by very large amounts of dietary cholesterol, therefore making the dietary induced hypercholesterolemia model less desirable for studies of the spontaneously occurring human disease. In our IHLC swine accelerated atherosclerosis occurs spontaneously when feeding low fat, cholesterol free diets. In conclusion, the IHLC pigs represent a unique and valuable resource as a model for studying human atherogenesis.

Current investigations on swine with inherited hyper-LDL-cholesterolemia (IHLC) and accelerated atherosclerosis

The development of a strain of swine with inherited LDL and cholesterol elevations (RAPACZ & HASLER-RAPACZ 1984), associated with accelerated atherosclerosis (RAPACZ et al. 1986b), stimulated further interest in characterization of LDL and its metabolism. Studies on chemical properties of low density lipoproteins (CHECOVICH et al. 1988), as well as on the profile and composition of lipoprotein subfractions from plasma of the IHLC swine (LEE et al. 1987a, LEE et al.1987b) showed that LDL exhibits a defective catabolism. Pigs with the Lpb5 mutant LDL's have a 30% lower catabolic rate *in vivo* and this accounts for hypercholesterolemia. This metabolic defect is associated with the larger, more buoyant cholesterol ester-rich lipoprotein particles than found in the LDL subfractions of normolipemic swine. It is conceivable that the accumulated larger more buoyant LDL in these pigs is especially atherogenic, leading spontaneously to the development of severe coronary atherosclerosis.

Since the IHLC pigs exhibit normal LDL receptor function, studies were undertaken to examine the activity of one of the major lipid enzymes: lecithin:cholesterol acyltransferase (LCAT). This enzyme regulates plasma cholesterol levels by catalyzing the formation of esterified cholesterol in the plasma of mammals in the presence of apolipoprotein AI (GLOMSET 1968). Completed studies on 41 IPH swine; 28 hypercholesterolemic and 13 normocholesterolemic showed that the LCAT activity was positively correlated with total cholesterol in the normocholesterolemic group (correlationcoefficient $r = +0.54$), whereas it was negatively correlated ($r = -0.82$) with total cholesterol in the hypercholesterolemic pigs. A further study revealed that the lower LCAT activity in hypercholesterolemic pigs is, at least in part, due to LCAT inhibition by components of the lipoprotein free fractions (LACKO et al., submitted). In the presence of LDL, LCAT carries another enzyme activity, lysolecithin acyltransferase (LAT) which acylates lysolecitin to lecithin. Free cholesterol to esterified cholesterol ratios and phospholipid sphingomyelin to lecitin ratios were significantly higher in plasma of IHLC pigs than in plasma of normolipemic pigs (SUBBAIAH et al., in preparation). It is likely that the defective LDL metabolism is the result of defective activities in LCAT and LAT.

Lipoprotein(a), Lp(a), is a LDL-like lipoprotein whose concentration in human plasma is associated with coronary atherosclerosis, however, its specific function and presence in nonprimate animal models are unknown. Plasminogen which is structurally very closely related to Lp(a) is the inactive plasma protein precursor of the proteolytic enzyme plasmin, which plays a central role in the fibrinolytic system and through this may be involved in coronary thrombosis, the most frequent cause of the fetal heart attack. Among identified polymorphic mutants in humans, a rare defective variant of plasminogen was found associated with recurrent thrombosis. Using anti-human plasminogen antibodies 11 different polymorphic forms were identified in 36 plasma samples derived from pigs of different Lpb genotypes (RAPACZ et al. 1989b). Studies on Lp(a) provide evidence for the presence in swine of lipoproteins having characteristics similar to Lp(a) found in humans and other primates and indicate that the pig may provide a model for studying the Lp(a) metabolism and function, and Apo(a) interactions with Lpb haplotypes (YE et al., submitted). Using anti-apoB epitope specific antibodies as immunosorbers, swine LDL from Lpb heterozygotes were separated recently into two apo-B haplotypes (RAPACZ et al. 1989a). These results confirmed previous studies on monogenic molecular expression of Lpb in LDL of Lpb heterozygous pigs. The existence of two heterozygous populations of LDL molecules will make possible studies on metabolic differences at a single *apo-B* gene level of two mutant Apo-B populations.

Acknowledgments

We appreciate the help of Mr. Buell Gunderson and Mr. Shu Cai Huang for animal care and treatment. We thank J. Busby for typing the manuscript. This research was supported in part by the College of Agricultural and Life Sciences and the Graduate School, University of Wisconsin, Madison, the Wisconsin Pork Producers, NIH AG 05856, and NIH HL 39774.

References

ALAUPOVIC, P. (1982): The role of apolipoproteins in lipid transport processes. La Ricerca Clin. Lab. 12, 1-21.
BARNES, R.H.; KWONG, E.; POND, W.; LOWRY, R.; LOOSLI, J.K. (1959): Dietary fat and protein and serum cholesterol. II. Young swine. J. Nutr. 69, 269-273.
BERG, K.; POWELL, L.M.; WALLIS, S.C.; PEASE, R.; KNOTT, T.J.; SCOTT, J. (1986): Genetic linkage between the antigenic group (Ag) variation and the apolipoprotein B gene: assignment of the Ag locus. Proc. Natl. Acad. Sci. USA 83, 7367-7370.
BRAGDON, J.H.; ZELLER, J.H.; STEVENSON, J.W. (1957): Swine and experimental atherosclerosis. Proc. Soc. Exp. Biol. Med. 95, 282-284.
BROWN, M.S.; GOLDSTEIN, J.S. (1986): A receptor-mediated pathway for cholesterol homeostatis. Science 232, 34-37.
BUSTAD, L.K.; McCLELLAN, R.O. (1966): Swine in Biomedical Research. Frayn Printing Co., Seattle.
BUTLER, R.; MORGANTI, G.; VIERUCCI, A. (1972): Serology and genetics of the Ag system. In: Proteides of the Biological Fluids (Editor: PEETERS, H.). Pergamon Press, Oxford. Vol. 19, Pp. 161-167.
CALLOWAY, D.H.; POTTS, R.B. (1962): Comparison of atherosclerosis in swine fed a human diet or purified diets. Circul. Res. 11, 47-52.
CHAPMAN, M.J. (1980): Animal lipoproteins: Chemistry, structure and comparative aspects. J. Lipid Res. 21, 789-853.
CHECOVICH, W.J.; FITCH, W.L.; KRAUSS, R.M.; SMITH, M.P.; RAPACZ J.; SMITH, C.L.; ATTIE, A.D. (1988): Defective catabolism and abnormal composition of low density lipoproteins from mutant pigs with hypercholesterolemia. Biochemistry 27, 1934-1941.
DOWNIE, H.G.; MUSTARD, J.F.; ROWSELL, H.C. (1963): Swine atherosclerosis: The relationship of lipids and blood coagulation to its development. Ann. N.Y. Acad. Sci. 104, 539-562.
FLESS, G.M.; ROLIH, C.A.; SCANU, A.M. (1984): Heterogeneity of human plasma lipoprotein(a). J. Biol. Chem. 259, 11470-11478.
FLORENTIN, R.A.; NAM, S.C.; DAOUD, A.S.; JONES, R.; SCOTT, R.F.; MORRISON, E.S.; KIM, D.N.; LEE, K.T.; THOMAS W.A.; DODD, W.J.; MILLER, K.D. (1968): Dietary-induced atherosclerosis in miniature swine. Parts I-V. Exp. Mol. Pathol. 8, 263-301.
FRENCH, J.E.; JENNINGS, M.A. (1965): The tunica intima of arteries of swine. In: Comparative Atherosclerosis (Editor: ROBERTS, J.C.). Harper and Row, New York. Pp. 25-36.
FRENCH, J.E.; JENNINGS, M.A.; POOLE, J.C.F.; ROBINSON, D.S.; Sir HOWARD FLOREY (1963): Intimal changes in the arteries of ageing swine. Proc. Roy. Soc. B, 158, 24-42.
GERRITY, R.G. (1981): The role of the monocyte in atherogenesis. I. Transition of blood-borne monocytes into foam cells in fatty lesions. Amer. J. Pathol. 103, 181-190.
GERRITY, R.G.; GOSS, J.A.; SOBY, L. (1985): Control of monocyte recruitment by chemotactic factor(s) in lesion-prone areas of swine aorta. Arteriosclerosis 5, 55-66.
GERRITY, R.G.; NAITO, H.K.; RICHARDSON, M.; SCHWARTZ, C.J. (1979): Dietary inducted atherogenesis in swine. Morphology of the intima in prelesion stages. Amer. J. Pathol. 95, 775-792.
GLOMSET, J.A. (1968): The plasma lecithin:cholesterol acyltransferase reaction. J. Lipid Res. 9, 155-167.
GOLDSMITH, D.P.J.; JACOBI, H.P. (1978): Atherogenesis in swine fed several types of lipid-cholesterol diets. Lipids 13, 174-189.
GOTTLIEB, H.; LALICH, J.J. (1954): The occurrence of arteriosclerosis in the aorta of swine. Amer. J. Pathol. 30, 851-855.
GREER, S.A.N.; HAYS, V.W.; SPEER, V.C.; McCALL, J.T. (1966): Effect of dietary fat, protein and cholesterol on atherosclerosis in swine. J. Nutr. 90, 183-190.
GRIGGS, T.R.; BAUMAN, R.W.; REDDICK, R.L.; READ, M.S.; KOCH, G.G.; LAMB, M.A. (1986): Development of coronary atherosclerosis in swine with severe hypercholesterolemia. Arteriosclerosis 6, 155-165.
HAMSTEN, A.; ISELIUS, L.; DAHLEN, G.; DE FAIRE, U. (1986): Genetic and cultural inheritance of serum lipids, low and high density lipoprotein cholesterol and serum apolipoproteins A-I, A-II and B. Atherosclerosis 60, 199-208.
HASLER-RAPACZ, J.; RAPACZ, J. (1982): Lipoprotein immunogenetics in primates. I. Two serum B-lipoprotein allotypes (Lmb1 and Lmb11) in rhesus monkeys and the LP-B immunological relationship with other primates. J. Med. Primatol. 11, 352-379.
HAVEL, R.J.; EDER, H.A.; BRAGDON, J.H. (1955): The distribution and chemical composition of ultracentrifugally separated lipoproteins in human serum. J. Clin. Invest. 34, 1345-1353.
HILL, E.G.; LUNDBERG, W.O.; TITUS, J.L. (1971): Experimental atherosclerosis. I. A comparison of menhaden-oil supplements in tallow and coconut-oil diets. Mayo Clin. Proc. 46, 613-620.
JACKSON, R.L. (1983): Plasma lipoproteins of swine: composition, structure and metabolism. In: CRC Handbook of Electrophoresis (Editor: LEWIS, L.A.). Vol. 4.

Jackson, R.L.; Baker, H.N.; Taunton, O.D.; Smith, L.C.; Garner, C.W.; Gotto, A.M., Jr. (1973): A comparison of the major apolipoprotein from pig and human high density lipoproteins. J. Biol. Chem. 248, 2639-2644.

Janado, M.; Martin, W.G. (1968): Molecular heterogeneity of a pig serum low-density lipoprotein. Can. J. Biochem. 46, 875-879.

Jennings, M.A.; Florey, H.W.; Stehbens, W.E.; French, J.E. (1961): Intimal changes in the arteries of a pig. J. Pathol. Bact. 81, 49-61.

Khan, M.A.; Earl, F.L.; Farber, T.M.; Miller, E.; Husain, M.M.; Nelson E.; Gertz, S.D.; Forbes, M.S.; Rennels, M.L.; Heald F.P. (1977): Elevation of serum cholesterol and increased fatty streaking in egg yolk:lard fed castrated miniature pigs. Exp. Mol. Pathol. 26, 63-74.

Kim, D.N.; Imai, H.; Schmee, J.; Lee, K.T.; Thomas, W.A. (1985): Intimal cell mass-derived atherosclerotic lesions in the abdominal aorta of hyperlipidemic swine. Atherosclerosis 56, 169-188.

Knipping, G.M.J.; Kostner, G.M.; Holasek, A. (1975): Studies on the composition of pig serum lipoproteins - isolation and characterization of different apoproteins. Biochim. Biophys. Acta 393, 88-99.

Lacko, A.G.; Lee, S.M.; Mirshahi, I.; Hasler-Rapacz, J.; Rapacz J.: Reduced LCAT activity in hypercholesterolemic pigs with apolipoprotein mutations. Atherosclerosis (submitted).

Lee, K.T. (1986): Cardiovascular. In: Swine in Biomedical Research (Editor: Tumbleson, M.E.). Plenum Press, New York. Vol. 3, Pp. 1481-1672.

Lee, D.M.; Mok T.; Hasler-Rapacz, J.; Rapacz, J. (1987a): Lipid and lipoprotein profiles in pigs with mutant apolipoprotein-B and hypercholesterolemia, homozygous and heterozygous. Feder. Proc. (FASEB) 46, 1340.

Lee, D.M.; Mok, T.; Hasler-Rapacz, J.; Rapacz, J. (1987b): Chemical composition of lipoprotein subfractions in swine with an apolipoprotein B mutation associated with hypercholesterolemia. Arteriosclerosis 7, 493a (Abstract).

Lee, K.T.; Jarmolych, J.; Kim, D.N.; Grant, C.; Krasney, J.A.; Thomas, W.A.; Bruno, A.M. (1971): Production of advanced coronary atherosclerosis, myocardial infarction and "sudden death" in swine. Exp. Mol. Pathol. 15, 170-190.

Lee, W.M.; Lee, K.T.; Thomas, W.A. (1979): Partial suppression by pyridinolcarbamate of growth and necrosis of atherosclerotic lesions in swine subjected to an atherogenic regimen that produces advanced lesions. Exp. Mol. Pathol. 30, 85-9.

Lewis, L.A.; Page, I.H. (1956): Hereditary obesity: relation to serum lipoproteins and protein concentrations in swine. Circulation XIV: 55-59.

Luginbühl, H.; Ratcliffe, H.L.; Detweiler, D.K. (1969): Failure of egg-yolk feeding to accelerate progress of atherosclerosis in older female swine. Virchows Arch. Abt. A. Pathol. Anat. 348, 281-289.

Ma, Y.; Schumaker, V.N.; Butler, R.; Sparkes, R.S. (1987): Two DNA restriction fragment length polymorphisms associated with Ag(t/z) and Ag(g/c) antigenic sites of human apolipoprotein B. Arteriosclerosis 7, 301-305.

Maeda, N.; Ebert, D.L.; Doers, T.M.; Newman, M.; Hasler-Rapacz, J.; Attie, A.D.; Rapacz, J.; Smithies, O. (1988): Molecular genetics of the apolipoprotein B gene in pigs in relation to atherosclerosis. Gene 70, 213-229.

Mahley, R.W. (1979a): Dietary fat, cholesterol and accelerated atherosclerosis. In: Atherosclerosis Reviews. (Editor: Paoletti, R.). Raven Press, New York, Vol. 5, Pp. 1-34.

Mahley, R.W. (1979b): Cholesterol-induced hyperlipoproteinemia and atherosclerosis in dogs, swine and monkeys: Models for human atherosclerosis. In: Atherosclerosis 5, Proc. Fifth Int. Symp. (Editor: Gotto, A.M.). Springer Verlag, New York. Pp. 355-358.

Mahley, R.W.; Innerarity, T.L.; Rall, S.C.; Weisgraber K.H. (1984): Plasma lipoproteins:apolipoprotein structure and function. J. Lipid Res. 25, 1277-1293.

Mahley, R.W.; Weisgraber, K.H. (1974): An electrophoretic method for the quantitative isolation of human and swine plasma lipoproteins. Biochemistry 13, 1964-1969.

Mahley, R.W.; Weisgraber, K.H.; Innerarity, T.; Brewer, H.B.,Jr.; Assmann, G. (1975): Swine lipoproteins and atherosclerosis. Changes in the plasma lipoproteins and apoproteins induced by cholesterol feeding. Biochemistry 14, 2817-2823.

McConathy, W.J.; Alaupovic, P. (1973): Isolation and partial characterization of apolipoprotein D: a new protein moiety of the human plasma lipoprotein system. FEBS Lett. 37, 178-182.

Moreland, A.F.; Clarkson ,T.B.; Lofland, H.B. (1963): Atherosclerosis in "Miniature" swine. I. Morphologic aspects. A. M. A. Arch. Pathol. 76, 203-210.

Pond, W.G.; Mersmann, H.J.; Young, L.D. (1986): Heritability of plasma cholesterol and triglyceride concentrations in swine. Proc. Exp. Biol. Med. 182, 221-224.

Pownall, H.J.; Jackson, R.L.; Roth, R.I.; Gotto, A.M.; Patsch, J.R.; Kummerow, F. A. (1980): Influence of an atherogenic diet on the structure of swine low density lipoproteins. J. Lipid Res. 21, 1108-1115.

RAPACZ, J. (1982): Current status of lipoprotein genetics applied to livestock production in swine and other domestic species. Proc. 2nd World Congress on Genetics Applied to Livestock Production (Graficas Orbe, Madrid), Vol. 4, Pp. 365-374.

RAPACZ, J. (1978): Lipoprotein immunogenetics and atherosclerosis. Amer. J. Med. Genet. 1, 377-405.

RAPACZ, J. (1974): Immunogenetic polymorphism and genetic control of low density β-lipoproteins in swine. Proc. First World Congress on Genetics Applied to Livestock Production (Graficas Orbe, Madrid), Vol. 1, Pp. 291-298.

RAPACZ, J.; ELSON, C.E.; LALICH, J.J. (1977): Correlation of an immunogenetically defined lipoprotein type with aortic intimal lipidosis in swine. Exp. Mol. Pathol. 27, 249-261.

RAPACZ, J.; HASLER-RAPACZ, J. (1989): Animal models - the pig. In: Genetic Factors in Atherosclerosis: Approaches and model systems (Editors: LUSIS, A.J.; SPARKES, R.S.). Monographs in Human Genetics Vol. 12, Karger, Basel. Pp. 139-169.

RAPACZ, J.; HASLER-RAPACZ, J. (1984): Investigations in the relationship between immunogenetic polymorphisms of β-lipoproteins and the β-lipoprotein and cholesterol levels in swine. In: Atherosclerosis and Cardiovascular Diseases. (Editors: LENZI, S.; DESCOVICH, G. C.). Editrice Compositori, Bologna. Pp. 99-108.

RAPACZ, J.; HASLER-RAPACZ, J.; KUO, W.H. (1986a): Immunogenetic polymorphism of lipoproteins in swine: genetic, immunological and physiochemical characterization of two allotypes Lpr1 and Lpr2. Genetics 113, 985-1007.

RAPACZ, J.; HASLER-RAPACZ, J.; KUO, W.H. (1978): Immunogenetic polymorphism of lipoproteins in swine. 2. Five new allotypic specificities (Lpp6, Lpp11, Lpp12, Lpp13 and Lpp14) in the Lpp system. Immunogenetics 6, 405-424.

RAPACZ, J.; HASLER-RAPACZ, J.; KUO, W.H.; LI D. (1976): Immunogenetic polymorphism of lipoproteins in swine. 1. Four additional serum lipoprotein allotypes (Lpp2, Lpp4, Lpp5 and Lpp15) in the Lpp system. Anim. Blood Groups Biochem. Genet. 7, 157-177.

RAPACZ, J.; HASLER-RAPACZ, J.; TAYLOR, K.M.; CHECOVICH, W.J.; ATTIE, A.D. (1986b): Lipoprotein mutations in pigs are associated with elevated plasma cholesterol and atherosclerosis. Science 234, 1573-1577.

RAPACZ, J.; HASLER-RAPACZ, J.; TAYLOR, K.M.; CHECOVICH, W.J.; ATTIE A.D. (1986c): Immunogenetically defined lipoprotein polymorphism associated with hypercholesterolemia and accelerated atherosclerosis in swine. 6th International Meeting "Atherosclerosis and Cardiovascular Diseases", Bologna, 318 (Abstracts).

RAPACZ, J., Jr.; HASLER-RAPACZ, J.; RAPACZ, J.; MCCONATHY, W.J. (1989a): Separation of swine plasma LDL from Lpb2/3 heterozygotes into two Apo-B allelic haplotypes, Lpb2 and Lpb3 with Apo-B epitope specific antibodies. J. Lipid Res. 30, 199-206.

RAPACZ, J.; GRUMMER, R.H.; HASLER, J.; SHACKELFORD, R.M. (1970): Allotype polymorphism of low density β-lipoproteins in pig serum (LDLpp1 and LDLpp2). Nature (London) 225, 941-942.

RAPACZ, J., Jr.; REINER, Z.; YE, S.Q.; HASLER-RAPACZ, J.; RAPACZ, J.; MCCONATHY, W.J. (1989b): Plasminogen polymorphism in swine. Compar. Biochem. Physiol. 93B, 325-331.

RATCLIFFE, H.L.; LUGINBÜHL, H.; PIVNIK, L. (1970): Coronary, aortic and cerebral atherosclerosis in swine of 3 age-groups: implications. Bull. Wld. Hlth. Org. 42, 225-234.

REISER, R.; SORRELS, M.F.; WILLIAMS, M.C. (1959): Influence of high levels of dietary fats and cholesterol on atherosclerosis and lipid distribution in swine. Circul. Res. 7, 833-846.

REITMAN, J.S.; MAHLEY, R.W. (1979): Changes induced in the lipoproteins of Yucatan miniature swine by cholesterol feeding. Biochim. Biophys. Acta 575, 446-457.

REITMAN, J.S.; MAHLEY, R.W.; FRY, D.L. (1982): Yucatan miniature swine as a model for diet-induced atherosclerosis. Atherosclerosis 43, 119-132.

ROBERTS, J.C., Jr.; STRAUS, R. (Editors) (1965): Comparative Atherosclerosis. Harper and Row, New York.

ROSENFELD, M.E.; PRESCOTT, M.F.; MCBRIDE, C.; RAPACZ, J.; HASLER-RAPACZ, J.; ATTIE, A.D. (1988): The composition and distribution of atherosclerotic lesions in the Lpb-5 spontaneously hypercholesterolemic pig closely resemble human atherosclerotic lesions. FASEB - Pathology (Abstract).

ROTHSCHILD, M.F.; CHAPMAN, A.B. (1976): Factors influencing serum cholesterol levels in swine. J. Hered. 67, 47-48.

ROWSELL, H.C.; DOWNIE, H.G.; MUSTARD, J.F. (1960): Comparison of the effect of egg yolk or butter on the development of atherosclerosis in swine. Can. Med. Assoc. J. 83, 1175-1186.

ROWSELL, H.C.; MUSTARD, J.F.; DOWNIE, H.G. (1965): Experimental atherosclerosis in swine. Ann. N.Y. Acad. Sci. 127, 742-762.

SCOTT, R.F.; KIM, D.N.; SCHMEE, J.; THOMAS, W.A. (1986a): Atherosclerotic lesions in coronary arteries of hyperlipidemic swine. Atherosclerosis 62, 1-10.

SCOTT, R.F.; REIDY, M.A.; KIM, D.N.; SCHMEE, J.; THOMAS, W.A. (1986b): Intimal cell mass derived atherosclerotic lesions in the abdominal aorta of hyperlipidemic swine. Atherosclerosis 62, 27-38.

SEGRETS, J.P.; ALBERS, J.J. (1986): Plasma lipoproteins, Part A; Methods in Enzymology, Vol. 128.

SKOLD, B.H.; GETTY R. (1961): Spontaneous atherosclerosis in swine. J. Amer. Vet. Med. Assoc. 139, 655-660.

St. Clair, R.W.; Bullock, B.C.; Lehner, N.D.M.; Clarkson, T.B.; Loftland, H.B., Jr. (1971): Long-term effects of dietary sucrose and starch on serum lipids and atherosclerosis in miniature swine. Exp. Mol. Pathol. **15**, 21-33.

Stunzi, H. (1965): Informal notes on arterial lesions in swine. Ann. N.Y. Acad. Sci. **127**, 740-742.

Thomas, W.A.; Kim, D.N.; Lee, K.T.; Reiner, J.M.; Scheme J. (1983): Population of dynamics of arterial cells during atherogenesis. Exp. Mol. Pathol. **39**, 257-270.

Watanabee, Y. (1980): Serial inbreeding of rabbits with hereditary hyperlipemia (WHHL-rabbit). Incidence and development of atherosclerosis. Atherosclerosis **36**, 261-268.

Ye, S.Q.; Rapacz, J., Jr.; Hasler-Rapacz, J.; Rapacz, J.; Stiers, D.; McConathy, W.J. (1989): Detection and isolation of swine lipoprotein(a). Biochim. Biophys. Acta (submitted).

Authors' address:
J. Rapacz and J. Hasler-Rapacz, University of Wisconsin, Immunogenetics Laboratory, 666 Animal Sciences Building, Madison, WI 53706, USA.

Caprine arthritis encephalitis

E. Peterhans, R. Zanoni, G. Ruff and S. Lazary

Introduction

The appreciation that AIDS in humans is caused by a member of the lentivirus subfamily (for a review see Gallo & Montagnier 1988) has led to an increase in research in animal diseases caused by related viruses. Among these diseases, caprine arthritis encephalitis in goats, Maedi Visna in sheep, infectious anemia in horses (for a review on these see Cheevers & McGuire 1988) and bovine Visna viruses (Gonda et al. 1987) have been at the center of interest. More recently, lentiviruses have also been reported in cats (Pedersen et al. 1987) and monkeys (Franchini et al., 1987 Chakrabarti et al. 1987). Caprine arthritis encephalitis (CAE) occurs worldwide. The incidence of infection with caprine arthritis encephalitis virus (CAE virus) may reach over 80% in some countries (Adams et al. 1984). The economic damage caused by CAE is primarily due to a gradual loss of performance (Smith & Cutlip 1988, Cheevers & McGuire 1988) manifested in decreased milk production and weight loss. Animals may have to be eliminated prematurely, thereby decreasing average productive lifespan in goat flocks. In the following, we shall discuss some basic properties of CAE virus and its relationship to Maedi Visna virus of sheep. This section will be followed by a description of the clinical course of infection with CAE virus and of the pathology resulting from the interaction between the virus and the host. Some aspects of the pathogenesis, epidemiology and diagnostic procedures will be described, along with observations that point to a role of the genetic setup of the animal in the development of disease. Finally, we shall focus on the potential use of CAE as a model for rheumatoid arthritis in humans and its use as a model to study certain aspects of lentiviral infection.

Caprine arthritis encephalitis virus

CAE virus is an enveloped RNA virus belonging to the lentivirinae subfamily of the retroviruses. It was first isolated by Crawford and coworkers (1980a) from goats suffering from arthritis. These authors proposed that the virus causing arthritis after intraarticular injection was identical with a virus that caused encephalitis in young kids and therefore named it "caprine arthritis encephalitis virus". Subsequent studies revealed that CAE virus was highly similar to the sheep lentivirus, Maedi Visna virus. Most of the proteins of CAE virus were found to induce antibodies crossreactive with Maedi Visna virus, allowing the serologic diagnosis of infection using antigen prepared from Maedi Visna virus (Houwers et al. 1982). Similar to other lentiviruses, CAE virus displays a considerable degree of "genetic plasticity". The virus may change over time within an infected animal thereby escaping immune surveillance (McGuire et al. 1988).

Clinical manifestations of infection with CAE virus

As appropriately described in the term "lenti", infections with these viruses are slow. This applies both to the duration of incubation and to the development of disease once the first symptoms have become evident. Infection with CAE virus may lead to at least three main symptoms. Arthritis is the most frequently observed symptom. An extensive clinical study has demonstrated the presence of arthritic lesions in approximately one third of serologically positive animals (Crawford & Adams 1981). Carpitis is the most easily observed manifestation of arthritis, but other articulations are also affected. Swelling of the carpal joints develops slowly over several months or even years and may reach a stage in which the periarticular region is affected, explaining why the disease is referred to as "big knee" by many goat owners (Fig.1). Obvious signs of pain are observed only late after the onset of carpitis. Taken together with the slow development of the change in shape of the carpal joint, this explains why some goat owners may overlook the most obvious sign of CAE for a long time. A second symptom of CAE, mastitis, may also escape detection. The milk remains unchanged and the

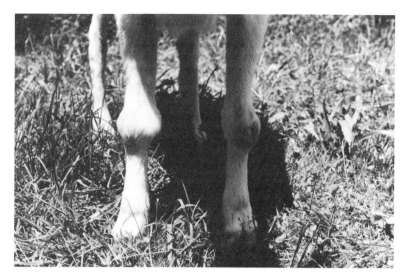

Fig. 1. Carpitis in a goat infected with CAE virus. Note marked swelling of carpal joints.

decrease in milk production in most cases is not marked enough or may develop over a more extended time period (ZWAHLEN et al. 1983, SMITH & CUTLIP 1988). In addition to arthritis and mastitis, CAE virus has also been reported to cause interstitial pneumonia. This manifestation of infection with CAE virus only rarely leads to clinically detectable symptoms (ROBINSON 1981, ELLIS et al. 1988). In young kids, the virus may cause encephalitis. Similar to Visna in sheep, but faster in its course, infection of the central nervous system begins with ataxia. This may be followed by paralysis first of the hind legs. Quadriplegia develops within 1-3 weeks after onset of disease symptoms. Typically, the animals do not develop fever and remain perceptive. Kids with mild disease may recover but may develop arthritis later in their lives (CRAWFORD et al. 1980b).

Pathology and pathogenesis

The histopathological signs of CAE resemble those of Maedi Visna, being characterized by a pronounced mononuclear inflammation in all affected organs. Although sometimes referred to as similar to multiple sclerosis, demyelination is not correlated with encephalitis caused by CAE virus (FANKHAUSER & THEUS 1983). The mechanism of life-long persistence is perhaps one of the most fascinating aspects of infection with CAE virus. Infection occurs in most cases through milk early in life. The site of primary viral multiplication is not known but may be in the lymphoreticular tissue of the nasopharyngeal region. The virus has a pronounced tropism for monocytes. These cells play an important role for the spread of the virus within the animal and virus can be isolated from peripheral blood monocytes by coculture with fibroblasts (PAWLISCH & MAES 1984). Several factors are believed to contribute to persistence of the virus, including limited genome expression, antigenic drift, and inefficient immune response. The limitation of genome expression in infected monocytes may prevent the detection of infected monocytes in the bloodstream by antiviral antibodies and T lymphocytes. Thus, as shown with the closely related Visna virus, monocytes produce little virus, whilst monocyte-derived macrophages are efficient host cells, correlating with a limited expression of viral genes in monocytes. Since monocytes mature into macrophages after emigrating from the bloodstream to the tissue, this situation has been referred to as "Trojan horse mechanism", indicating that monocytes enter the tissue without being detected as carriers of the virus (PELUSO et al. 1985). Once viral multiplication is initiated in the macrophages, the host produces a lentivirus-specific type of interferon which limits the extent of viral multiplication by slowing down the maturation of monocytes and inhibiting viral replication in macrophages. In addition, other cytokines may be released from the mononuclear cells present at the site of viral multiplication, leading to the establishment of the typical chronic inflammatory response observed in lentivirus-infected goats and sheep (for a review see ZINK et al. 1987). It was long believed that sheep, but not goats develop neutralizing antiviral antibodies to Maedi Visna and CAE viruses, respectively. However, neutralizing antibodies have recently been demonstrated also in goats (ELLIS et al. 1987, McGUIRE et al. 1988). The titers demonstrated in these animals were low, however. In addition, since neutralization-sensitive epitopes were found to change in the course of persistent infection, neutralizing antibodies represent an inefficient defense mechanism in CAE (McGUIRE et al. 1988).

Another interesting aspect of CAE is the way in which the virus causes damage in the infected animal. Clearly, CAE virus causes no immunosuppression in infected animals (DeMARTINI et al. 1983). There is circumstantial evidence that the antiviral immune response may bring about detrimental effects for the animal. Thus, immunization of goats with inactivated virus increased the severity of arthritis (McGURIE 1987) whilst drug-induced immunosuppression decreased the histopathological lesions in Visna (NATHANSON et al. 1976). In Visna, no evidence was found for a possible contribution of autoimmune mechanisms. We have recently proposed that part of the pathology may be caused by "autotoxicity" (PETERHANS et al. 1988). This mechanism is different from autoimmunity by lacking immunological focus on "self" (i.e. autoantibody, self-reactive T cells) but implies that the virus either directly activates effector mechanisms such as the generation of oxygen-derived free radicals from phagocytic cells, or, during multiplication in monocyte/macrophages, activates host genes coding for the production of monokines such as tumor necrosis factor (TNF) and interleukin-1. In fact, recombinant human TNF-α has recently been demonstrated to stimulate the resorption of cartilage and inhibit the synthesis of proteoglycans in explants of cartilage (SAKLATVALA 1986). Another cytokine, Interferon, has recently been demonstrated to increase in concentration in synovial fluid of goats suffering from arthritis (YILMA et al. 1988). This finding may be significant with respect to the proposed contribution of TNF to arthritis caused by CAE virus because interferons are known to enhance TNF-mediated monocyte cytotoxicity (PHILIP & EPSTEIN 1986).

Epidemiology

Colostrum and milk of infected goats are considered the most important means of viral transmission (CRAWFORD et al. 1980a, ELLIS et al. 1983). This view is supported by reports of successful virus isolation from goat milk (ADAMS et al. 1983b, ELLIS et al. 1983). Horizontal transmission by other means seems to contribute to the spread of infection between animals kept in close contact (ADAMS et al. 1983a, ADAMS et al. 1983b). Although virus has been demonstrated in semen of infected bucks, sexual transmission is probably of limited importance (ADAMS et al. 1983b). Detailed studies on intrauterine virus transmission are lacking, however, epidemiological observations made in kids separated from their mothers immediately after birth and fed cow's colostrum argue against an important role of this mode of transmission. The question of whether Maedi Visna virus can be transmitted to goats, and CAE virus to sheep, is of interest both for scientific and practical reasons. While interspecies transmission in both directions has been demonstrated experimentally (OLIVER et al. 1982, BANKS et al.1983, SMITH et al. 1985), there is no convincing evidence that this occurs also under field conditions. In particular, the incidence of Maedi Visna and CAE may be quite different. For example, the incidence of Maedi Visna in Switzerland is approximately 4% whilst that of CAE is over 60% (KRIEG 1989).

Diagnosis of CAE

The diagnosis of CAE is based mainly on the detection of antiviral antibodies by agar gel diffusion, indirect ELISA or complex trapping blocking assays (ADAMS et al. 1985, HOUWERS et al. 1982, HOUWERS & SCHAAKE 1987, COACKLEY et al. 1984). Western blotting may be used as a backup method for antibody detection in cases which do not give clear results with the other tests (ZANONI et al. 1989). In contrast to serology, virus isolation is not normally used to diagnose infection with CAE virus, mainly because of the long time required for the initial isolation. Furthermore, antisera with a high concentration of neutralizing antibody are not readily available, making unequivocal identification of CAE virus difficult. We have recently demonstrated that Western blotting allows the demonstration of CAE viral proteins in the supernatant of cell cultures within 5 days post-infection (ZANONI et al. 1989, ZANONI et al., submitted) . This technique may be useful in selected cases, e.g. in time course studies of experimental infection. An alternative approach aims at the demonstration of viral RNA or DNA. Using in situ hybridization techniques, HAASE and coworkers analysed the kinetics of viral DNA and RNA synthesis and demonstrated a considerable heterogeneity in the speed of replication between individual cells in a cell culture (HAASE et al. 1982). Although highly sensitive in the detection of viral DNA and RNA, *in situ* hybridization takes several days to weeks to perform. We have recently explored the possibility of demonstrating viral DNA using the polymerase chain reaction. This novel technique is capable of amplifying by up to 10^6 fold specific sequences of DNA and takes only some 8 hours to perform (SAIKI et al. 1985). Using this technique, we were able to demonstrate DNA of CAE virus as early as 1 day post-infection in cultured cells (ZANONI et al., submitted). If successfully adapted to material such as mononuclear cells obtained from infected animals, the polymerase chain reaction could be extremely useful in studies on the pathogenesis of CAE.

Control of CAE

As described in the introduction, CAE occurs worldwide and the incidence of infection is quite high in some countries. In Switzerland for example, over 60% of the goats are infected (KRIEG 1989). Control and eventual eradication programs face also other problems. Thus, it may take several weeks before antiviral antibodies can be detected in infected animals. Furthermore, disease symptoms develop slowly and are seldom very obvious which may explain why many goat owners get used to CAE and tend to view it as a minor problem of little economic consequence. Similar to AIDS and other lentiviral infections, vaccines are not available and no therapy exists that would terminate the disease. Based on these aspects and taking into account the pathogenesis and epidemiology of CAE, similar control programs have been initiated in several countries. The most important measure consists in the separation of newborn kids from their mothers and rearing with CAE-free colostrum and milk. This can conveniently be done using cow's colostrum and milk (ELLIS et al. 1983, HOUWERS et al. 1983, BALCER et al. 1985). The state of the flocks is controlled by serologic testing for antibodies to CAE virus. Care must be taken to avoid contact with infected animals because of possible horizontal transmission of the virus. This method of control of CAE has the advantage of preserving valuable genetic material, however, it depends heavily on the cooperation of the goat owners. In our experience, it is possible to obtain CAE-free flocks if the goat owners strictly adhere to the measures.

Genetic influence on disease development

In Switzerland, BÜCHI et al. (1962) described arthritis of unknown etiology in the Toggenburg goat and spoke of genetic-environmental interactionsnv influencing disease development. The putative agent has later been demonstrated to be of retrovirus nature as described in the preceeding sections. The predisposition to development of disease symptoms observed in certain lines has also been noted in other Swiss breeds and was further investigated for its underlying immunogenetic mechanisms.

In order to clarify a possible involvement of the major histocompatibility complex (MHC) in the disease pathogenesis of caprine arthritis, the caprine leukocyte antigen (CLA) system has been serologically characterized (RUFF 1987, NESSE & RUFF 1989) and was used as a marker system for immunological traits in family and population studies.

All investigated animals were infected with CAE virus, showing a humoral response to CAE virus in the ELISA test. When comparing affected and healthy animals, differences in the distribution of CLA class I antigen frequencies were not found at the population level, except for one of the antigens

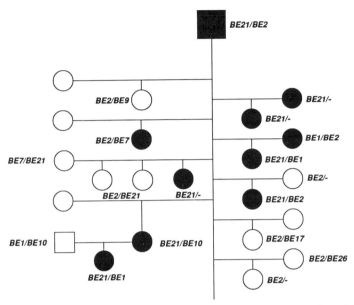

Fig. 2. *Inheritance of the paternal class I-marked haplotypes in CAE-affected (black) and healthy (white) progeny of a Saanen sire. The ratio of the antigens BE21 to BE2 was 4:1 in the informative affected and 0.5 in the healthy group. Squares represent male goats, and circles represent female goats. (from* RUFF & LAZARY *1988).*

in the Saanen breed, where carriers seemed to be less prone to develop arthritis symptoms than animals lacking this antigen. Eleven groups of comparable environment (multiple-case families or halfsibling groups with at least two informative diseased offspring/group) were analyzed for manifestation of the disease and segregation of the parental CLA class I marked haplotypes (e.g. in Fig.2). The antigens were inherited in a ratio of 1:1 in all the offspring tested, but showed a significant segregation distorsion when classifying the progeny in affected and healthy groups. The statistical analysis gave evidence for linkage between the locus encoding the determined class I alleles and a hypothetical locus (i) coding for genes responsible for arthritis resistance/susceptibility (RUFF & LAZARY 1988).

These data provided the first evidence that CAE virus-induced arthritis in the goat is genetically influenced by the MHC region. They also suggested that susceptibility/resistance was not directly associated with the determined class I gene products but rather is in close genetic linkage as the particular class I CLA allele associated with the disease susceptibility varied from family to family.

CAE as a model for biomedical research.

The pathogenesis of human rheumatoid arthritis (CHRISTIAN & PAGET 1987) has remained unclear ever since the first description of the disease. Many causative agents, i.e., viral, bacterial and chemical, have been suspected (BENNETT 1978). One of the difficulties in defining the etiology and pathogenesis of the disease has been the lack of a spontaneous model.

Several experimentally induced arthritides, which display one or more of the features of rheumatoid arthritis (RA) have been investigated. Type II collagen induced arthritis in rats (TRENTHAM et al. 1977) and in mice (COURTENAY et al. 1980); mycoplasma-induced arthritis in rabbits (COLE et al. 1977) and in mice (TAYLOR et al. 1978) or the use of streptococcal cells and cell walls as arthritis inducing agents in rats (CROMARTIE et al. 1977) may serve as examples. Induction of arthritis by injection of synthetic adjuvant in rats has been described by CHANG and coworkers (1980).

In 1959, SIKES reported a rheumatoid-like arthritis in swine caused by *Erysipelothrix insidiosa* (SIKES 1959). VAN PELT (1965) described cattle with malignant lymphoma suffering from degenerative joint disease. A spontaneously occurring, arthritis-like disease found in mice has been described by LeMing and coworkers (1982), very similar to human RA, showing increased levels of circulating IgM rheumatoid factor in close correlation to demonstrable synovitis and/or arthritis.

In contrast to the human and murine model, rheumatoid factors and immune complexes have not been detected in the goat model (Zwahlen et al. 1985). With the putative agent being a representative of the retrovirus group, this animal model of spontaneous occurring arthritis is of great interest for comparative studies. The hypothesis of the involvement of a retrovirus in the pathogenesis of RA has not been ruled out. Furthermore, susceptibility to collagen-induced arthritis in rats (Griffiths et al. 1981) and mice (Wooley et al. 1981) as well as to human rheumatoid arthritis (Svejgaard et al. 1983) is apparently associated with the major histocompatibility complex (MHC).

Our findings concerning the influence of the goat MHC or gene(s) associated with it on disease development as well as the retrovirus etiology in disease pathogenesis make this natural occurring form of arthritis an interesting model for basic research studies. The application of molecular genetics in comparison with investigations at the product level should elucidate the influence of the MHC and gene(s) linked to it in disease pathogenesis and may also proof to be of valuable help for better understanding of disease pathogenesis in human rheumathoid arthritis.

The interaction between CAE virus and the goat is of comparative interest with respect to lentiviral infections of humans. Although not leading to immunodeficiency, CAE virus persists in its host and chronic disease is observed after long and variable incubation time. Using mononuclear cells derived from genetically defined goats, it will be possible to study virus-host interaction at the cellular and molecular level and to elucidate the potential role of host genetics in the outcome of this interaction.

Acknowledgements

This work was supported by Swiss National Science Foundation grants 3.636-0.87 to EP and 3.879-0.88 to SL.

References

Adams, D.S.; Allonby, M.E.W.; Bell, J.F.; Waghela, S.; Heinonen, R. (1983a): Observations on caprine arthritis-encephalitis in Kenya. The Veterinary Record 112, 227-228.

Adams, D.S.; Gogolewski, R.P.; Barbet, A.F.; Cheevers, W.P. (1985): Identification of caprine arthritis-encephalitis retrovirus proteins in immunodiffusion precipitin lines. J. Gen. Virol. 66, 1139-1143.

Adams, D.S.; Klevjer-Anderson, P.; Carlson, J.L.; McGuire T.C. (1983b): Transmission and control of caprine arthritis-encephalitis virus. Amer. J. Vet. Res. 44, 1670-1675.

Adams, D.S.; Oliver, R.E.; Ameghino, E.; DeMartini, J.C.; Verwoerd, W.; Houwers, D.J.; Waghela, S.; Gorham J.R.; Hyllseth, B.; Dawson, M.; Trigo F.J. (1984): Global survey of serological evidence of caprine arthritis-encephalitis virus infection. Vet. Rec. 115, 493-495.

Balcer, T.; Stucki, M.; Krieg, A.; Zwahlen R. (1985): Caprine retrovirus infection: experience with a pilot sanitation program in Swiss goat herds. In: Slow Viruses in Sheep, Goats and Cattle (Editors: Sharp, J. M.; Hoff-Jorgensen, R.). Commission of the European Communities, Brussels and Luxembourg. Pp. 253-264.

Banks, K.L.; Adams, D.S.; McGuire, T.C.; Carlson, J. (1983): Experimental infection of sheep by caprine arthritis-ecephalitis virus and goats by progressive pneumonia virus. Amer. J. Vet. Res. 44, 2307-2311.

Bennett, J.C. (1978): The infectious etiology of rheumatoid arthritis. New considerations. Arth. Rheum. 21, 531-538.

Büchi, H.F.; Le Roy, H.L.; Böni, A. (1962): Über das Auftreten von Polyarthritis bei der Species *Capra hircus*, ein medizinisch-genetisches Problem. Zeitschrift f. Rheumaforschung, 21, 88-09.

Chakrabarti, L.; Guyader, M.; Alizon, M.; Daniel, M.D.; Desrosiers, RC.; Tiollais, P.; Sonigo, P. (1987): Sequence of simian immunodeficiency virus from macaque and its relationship to other human and simian retroviruses. Nature 328, 543-547.

Chang, Y.; Pearson, C.M.; Abe, C. (1980): Adjuvant polyarthritis. IV. Induction by a synthetic adjuvant: immunologic, histopathologic and other studies. Arthritis. Rheum. 23, 62.

Cheevers, W.P.; McGuire, T.C. (1988): The lentiviruses: Maedi/Visna, caprine arthritis-encephalitis, and equine infectious anemia. In: Advances of Virus Research, Vol. 34. (Editors: Maramorosch, K.; Murphy, F. A.; Shatkin, A. J.). Academic Press, Orlando, USA. Pp. 189-215.

Christian, C.L.; Paget, S.A. (1987): Rheumatoid arthritis. In: Immunological diseases, 3rd edition. (Editor: Samter, M.). Little, Brown and Co., Boston. Pp. 1061-1065.

COACKLEY, W.; SMITH, V.W.; HOUWERS, D.J. (1984): Preparation and evaluation of antigens used in serological tests for caprine syncytial retrovirus antibody in sheep and goat sera. Vet. Microbiol. **9**, 581-586.

COLE, B.C.; GRIFFITHS, M.M.; EICHWALD, E.J.; WARD, J.R. (1977): New models of chronic synovitis in rabbits induced by mycoplasmas: microbiological, histopathological and immunological observation on rabbits injected with *Mycoplasma arthritidis* and *Mycoplasma pulmonis*. Infec. Immunity. **16**, 383-396.

COURTENAY, J.S.; DALLMAN, M.J.; DAYAN, A.D.; MARTIN, A.; MOSEDAL, B. (1980): Immunisation against heterologous type II collagen induces arthritis in mice. Nature **283**, 666-668.

CRAWFORD, T.B.; ADAMS, D.S. (1981): Caprine arthritis-encephalitis: clinical features and presence of antibody in selected goat populations. Amer. Vet. Med. Assoc. **178**, 713-719.

CRAWFORD, T.B.; ADAMS, D.S.; CHEEVERS, W.P.; CORK, L.V. (1980a): Chronic arthritis in goats caused by a retrovirus. Science **207**, 997-999.

CRAWFORD, T.B.; ADAMS, D.S.; SANDE, R.D.; GORHAM, J.R.; Henson J.B. (1980b): The connective tissue component of the caprine arthritis-encephalitis syndrome. Amer. J. Pathol. **100**, 443-454.

CROMARTIE, W.J.; CRADDOCK, J.G.; SCHWAB, J.H.; ANDERLE, S.K.; YANG, C.H. (1977): Arthritis in rats after systemic injection of streptococcal cells or cell walls. J.Exp.Med. **146**, 1585-1602.

DEMARTINI, J.C.; BANKS, K.L.; GREENLEE, A.; ADAMS, D.S. (1983): Augmented T lymphocyte responses and abnormal B lymphocyte numbers in goats chronically infected with the retrovirus causing caprine arthritis-encephalitis. Amer. J. Vet. Res. **44**, 2064-2069.

ELLIS, T.M.; ROBINSON, W.F.; WILCOX, G.E. (1988): The pathology and aetiology of lung lesions in goats infected with caprine arthritis-encephalitis virus. Aust. Vet. J. **65**, 69-73.

ELLIS, T.; ROBINSON, W.; WILCOX G. (1983): Effect of colostrum deprivation of goat kids on the natural transmission of caprine retrovirus infection. Aust. Vet. J. **60**, 326-329.

ELLIS, T.M.; WILCOX, G.E.; ROBINSON, W.F. (1987): Antigenic variation of Caprine Arthritis-Encephalitis Virus during persistent infection of goats. J. Gen.Virol. **68**, 3145-3152.

FANKHAUSER, R.; THEUS, T. (1983): Visna bei der Ziege. Schw. Arch. Tierheilk. **125**, 387-390.

FRANCHINI, G.; GURGO, C.; GUO, H.G.; GALLO, R.C.; COLLALTI, E., FARGNOLI, K.A.; HALL, L.F.; WONG-STAAL, F.; REITZ, M.S., Jr. (1987): Sequence of simian immunodeficiency virus and its relationship to the human immunodeficiency viruses. Nature **328**, 539-543.

GALLO, R.C.; MONTAGNIER, L. (1988): AIDS in 1988. Sci. Amer. **259**, 25-32.

GONDA, M.S.; BRAUN, M.J.; CARTER, S.G.; KOST, T.A.; BESS, J.W., Jr.; ARTHUR, L.O.; VAN DER MAATEN, M.J. (1987): Characterization and molecular cloning of a bovine lentivirus related to human immunodeficiency virus. Nature **370**, 388-391.

GRIFFITHS, M.M.; EICHWALD, E.J.; MARTIN, M.H.; SMITH, C.B.; DEWITT, C.W. (1981): Immunogenetic control of experimental type II collagen induced arthritis. I. Susceptibility and resistance among inbred strains of rats. Arth. Rheum. **24**, 781-789.

HAASE, A.T.; STOWRING, L.; HARRIS, J.D.; TRAYNOR, B.; VENTURA, P.; PELUSO, R.; BRAHIC, M. (1982): Visna DNA synthesis and the tempo of infection *in vitro*. Virology **119**, 399-410.

HOUWERS, D.J.; GIELKENS, A.L.J.; SCHAAKE, J. (1982): An indirect enzyme-linked immunosorbent assay (ELISA) for the detection of antibodies to maedi-visna virus. Vet. Microbiol. **7**, 209-219.

HOUWERS, D.J.; KOENIG, C.D.; DE BOER, G.F.; SCHAAKE, J.Jr (1983): Maedi-visna control in sheep. I. Artificial rearing of colostrum-deprived lambs. Vet. Microbiol. **8**, 179-185.

HOUWERS, D.J.; SCHAAKE, J., Jr. (1987): An improved ELISA for the detection of antibodies to ovine and caprine lentiviruses, employing monoclonal antibodies in a one-step assay. J. Immunol. Meth. **98**, 151-154.

KRIEG, A. Die Caprine-Arthrithis-Enzephalitis in der Schweiz: Serologische und klinische Untersuchungen. Inaugural Dissertation der Vet. Med. Fakultät der Universität Bern. (Thesis).

LEMING, H.; THEOFILOPOULOS, A.N.; DIXON, F.J. (1982): A spontaneous rheumatoid arthritis-like disease in MRL/1 mice. J.Exp.Med. **155**, 1690-1701.

MCGUIRE, T.C. (1987): The immune response to viral antigens as a determinat of arthritis in capine arthritis-encephalitis virus infection. Vet. Immun. Immunopathol. **17**, 465-470

MCGUIRE, T.C.; NORTON, L.K.; O'ROURKE, K.I.; CHEEVERS, W.P. (1988): Antigenic variation of neutralization-sensitive epitopes of caprine arthritis-encephalitis lentivirus during persistent arthritis. J.Virol. **62**, 3488-3492.

NATHANSON, N.; PANITCH, H.; PALSSON, P.A.; PETURSSON, G.; GEORGSSON, G. (1976): Pathogenesis of visna. II. Effect of immunosuppression upon early central nervous system lesions. Lab. Invest. **35**, 444-460.

NESSE, L.L.; RUFF, G. (1989): A comparison of lymphocyte antigen specificities in Norwegian and Swiss goats. Anim. Genet. **20**, 33-39.

OLIVER, R.E.; MCNIVEN, R.A.; JULIAN, A.F. (1982): Experimental infection of sheep and goats with caprine arthritis-encephalitis virus. New Zealand Vet. J. **30**, 158-159.

PAWLISCH, R.A.; MAES, R.K. (1984): Caprine arthritis-encephalitis virus isolated from Michigan goats. Amer. J. Vet. Res. **45**, 1808-1811.

PEDERSEN, N.C.; HO, E.W.; BROWN, M.L.; YAMAMOTO, J.K. (1987): Isolation of a T-lymphotropic virus from domestic cats with an immunodeficiency-like syndrome. Science 235, 790-793.

PELUSO, R.; HAASE, A.; STOWRING, L.; EDWARDS, M.; Ventura, P. (1985): A Trojan horse mechanism for the spread of Visna virus in monocytes. Virology 147, 231-236.

PETERHANS, E.; JUNGI, T.J.; STOCKER, R. (1988): Autotoxicity and reactive oxygen in viral disease. In: Oxy-Radicals in Molecular and Cellular Biology. New Series, Vol. 82. (Editors: CERUTTI, P.; FRIDOVICH, I.; McCord, J.). Alan R. Liss Inc., N.Y.. Pp. 543-562.

PHILIP, R.; EPSTEIN, L.B. (1986): Tumour necrosis factor as immunomodulator and mediator of monocyte cytotoxicity by itself, gamma-interferon and interleukin-1. Nature 323, 86-89.

ROBINSON, W.F. (1981): Chronic interstitial pneumonia in association with a granulomatous encephalitis in a goat. Aust. Vet. J. 57, 127-131.

RUFF, G. (1987): Investigations on the caprine leucocyte antigen (CLA) system. Thesis No. 8468 ETH, Zürich.

RUFF, G.; LAZARY, S. (1988): Evidence for linkage between the caprine leucocyte antigen (CLA) system and susceptibility to CAE virus-induced arthritis in goats. Immunogenetics 28, 303-309.

SAIKI, R.K.; SCHARF, S.; FALOONA, F.; MULLIS, K.B.; HORN, G.T.; ERLICH, H.A.; ARNHEIM N. (1985): Enzymatic amplification of ß-globulin genomic sequences and restriction site analysis for diagnosis of sickle cell anemia. Science 230, 1350-1354.

SAKLATVALA, J. (1986): Tumour necrosis factor-α stimulates resorption and inhibits synthesis of proteoglycan in cartilage. Nature 322, 547-549.

SIKES, D. (1959): A rheumatoidlike arthritis in swine. Lab. Invest. 8, 1406-1415.

SMITH, M.C.; CUTLIP, R. (1988): Effects of infection with caprine arthritis-encephalitis virus on milk production in goats. J. Amer. Vet. Med. Assoc. 193, 63-67.

SMITH, V.W.; DICKSON, J.; COACKLEY, W.; CARMAN, H. (1985): Response of merino sheep to inoculation with a caprine retrovirus. Vet. Rec. 117, 61-63.

SVEJGAARD, A.; PLATZ, P.; RYDER, L.P. (1983): HLA and disease 1982 - A survey. Immun.Rev. 70, 193-218.

TAYLOR, G.; TAYLOR-ROBINSON, D.; KEYSTONE, E.C. (1978): Effects of lymecycline on *Mycoplasma pulmonis*-induced arthritis in mice. Brit. J. Exp. Pathol. 59, 204.

TRENTHAM, D.E.; TOWNES, A.S.; KANG, A.H. (1977): Autoimmunity to type II collagen: an experimental model of arthritis. J. Exp. Med. 146, 857.

VAN PELT, R.W. (1965): Comparative arthology in man and domestic animals. J. Amer. Vet. Med. Assoc. 147, 958-967.

WOOLEY, P.H.; LUTHRA, H.S.; STUART, J.M.; DAVID, C.S. (1981): Type II collagen induced arthritis in mice I. Major histocompatibility complex (I region) linkage and antibody correlates. J. Exp. Med. 154, 688-700.

YILMA, T.; OWENS, S.; ADAMS, D.S. (1988): High levels of interferon in synovial fluid of retrovirus-infected goats. J.Interferon Res. 8, 45-50.

ZANONI, R.; KRIEG, A.; PETERHANS, E. (1989): Detection of antibodies to caprine arthritis-encephalitis virus by protein G enzyme-linked immunosorbent assay and immunoblotting. J. Clin. Microbiol. 27, 580-582.

ZINK, M.; NARAYAN, O.; KENNEDY, P.G.E.; CLEMENTS, J.E. (1987): Pathogenesis of Visna/Maedi and caprine arthritis-encephalitis: new leads on the mechanism of restricted virus replication and persistent infection. Vet. Immunol. Immunopathol. 15, 167-180.

ZWAHLEN, R.; AESCHBACHER, M.; BALCER, TH.; STUCKI, M.; WYDER-WALTHER, M.; WEISS, M.; STECK, F. (1983): Lentivirusinfektionen bei Ziegen mit Carpitis und interstitieller Mastitis. Schweiz. Arch. Tierheilk. 125, 281-299.

ZWAHLEN, R.; SPAETH, P.J.; STUCKI, M. (1985): Histological and immunopathological investigations in goats with carpitis/pericarpitis. In: Slow Viruses in Sheep, Goats and Cattle. (Editors: SHARP, J. M.; HOFF-JORGENSEN, R.). Commission of the European Communities, Brussels and Luxembourg. Pp. 239-248.

Authors' adresses:
E. Peterhans and R. Zanoni, Institute of Veterinary Virology, University of Berne, Länggass-Str. 122, CH-3012 Bern, Switzerland;
G. Ruff and S. Lazary, Institute of Animal Husbandry, University of Berne, Länggass-Str. 122, CH- 3012 Bern, Switzerland.

Hemophilia in sheep and the use of sheep in blood coagulation research

S. NEUENSCHWANDER and V. PLIŠKA

Physiological control of bleeding: hemostasis

Bleeding incidents (hemorrhage) are widespread in men and animals. They occur either after injuries or as a consequence of pathogenic circumstances, such as rupture of a sclerotic blood vessel, lesions or dysfunctions of the coagulation system. They can attain different stages of severity. When bleeding is not immediately controlled by physiological mechanisms (hemostasis) or therapeutic means, it may result in critical loss of body fluids within a short period of time. Blood clotting processes, the primary defense against hemorrhage, result in an instantaneous, mechanical block of the blood outflow, provided that this occurs in a small vessel and under a low blood pressure. Humoral control of blood pressure and of water balance (redistribution of body fluids) in higher vertebrates plays a supporting role in the preservation of fluids in the event of severe bleeding, which may lead to a fatal hemorrhagic shock. Wound healing processes eventually accomplish the repair of the damaged site.

The primary defense against bleeding is comprised of several steps (Fig. 1). Immediately after injury platelets from the outflowing blood adhere to the collagen of the damaged vessel wall and release serotonin, a vasoactive endogenous amine which elicits the contraction of the smooth muscle fibres (Fig. 1a). Although capillaries do not contain muscle fibres, blood loss is prevented by endothelial cell adhesion and by the response of subendothelial myofibrillar structures. Other materials released by the platelets, like adenosine diphosphate (ADP), thromboxanes, calcium ions, chemotactic factors and platelet factors, amplify the adhesion of platelets (Fig. 1b). Finally, the wound is filled with fibrin which stabilizes the formed platelet plug (Fig. 1c).

Blood coagulation represents a cascade of various enzymatic processes. Inactive proenzymes (coagulation factors) are activated, initiating the next step (MacFarlane 1964, Davie & Ratnoff 1964). Three principle steps in the formation of a blood clot may be distinguished:
1) formation of activated factor X (F Xa);
2) formation of thrombin;
3) formation of fibrin.

The sources of the compounds needed for these essential steps are found in tissues, plasma and platelets (Fig. 2). Thromboplastin from the damaged vessel walls, and from the platelets, activates F VII, and the contact of plasma with a negatively charged surface results in activation of F XI and F XII. Activated F IX together with F VIII, calcium ions and phospholipids are potent F X activators which catalyze the generation of thrombin. Finally, thrombin converts fibrinogen to an insoluble fibrin matrix by proteolytic cleavage.

Human hemophilia

As in any other complex system, a partial dysfunction of individual steps can be repaired by compensation through the remaining processes. A total failure at a certain level of the coagulation cascade, however, usually results in a severe pathological condition, called a coagulopathy. The most common causes of coagulopathies are absence of active coagulation factors (usually because of a genetic defect), reduced number of platelets (thrombocytopenia), disturbed platelet function, or vessel wall defects (von Willebrand's disease).

Pathology

A lack of factors VIII or IX, or a mutated coagulation factor, results in an insufficient thrombin generation. As a consequence, the formed platelet plug fails to be stabilized, and the bleeding continues. Thus, the disease is manifested by intermittent bleeding, spontaneous or after small injuries,

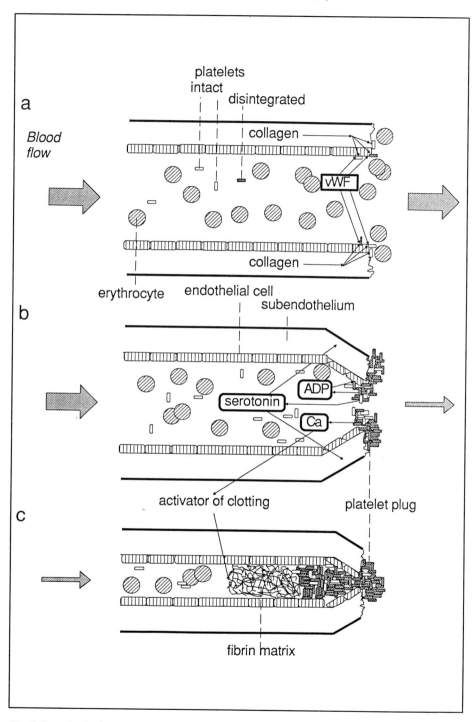

Fig. 1. Repair of a damaged blood vessel showing different stages of the complex mechanisms: a) injury, b) reduction of the wound size, and c) sealing of the vessel with a fibrin plug.

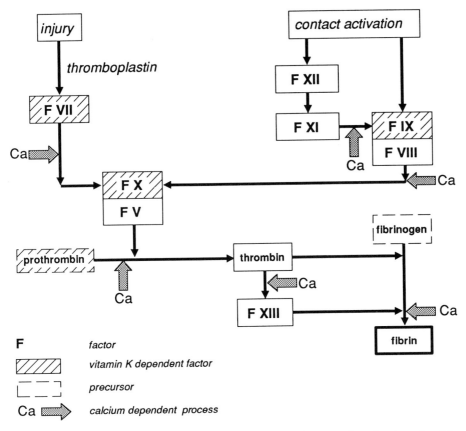

Fig. 2. Main activation pathways of the coagulation cascade. Control mechanisms involving inhibitors are not shown.

mostly in extremities and in subserosal tissue. "After-bleeding incidents" can occur in more severe forms of the disease. Typical for this condition is, also, bleeding into joints which leads to hemarthroses and may seriously hinder the movements of limbs.

The severity of hemophilia A is closely related to the *in vitro* estimated plasma F VIII clotting activity. In mild or moderate forms (subhemophilia), with an F VIII activity of more than 25% of the norm, complications generally occur only after surgery or after major injuries. Forms with 5% or less F VIII activity are clinically classified as severe and may lead to spontaneous, life-threatening bleeding.

History

The earliest recorded account of the complications of minor surgery performed on hemophiliacs is found in the fifth century in the Babylonian Talmud (ROSNER 1969). At that time, the third child was not permitted to be circumcised if the first two had died as a result of that procedure. In the nineteenth century, the first reference appeared to the oldest and largest hemophilia family ever described: the Tenna family, named after a village in Switzerland where generations have lived for about seven centuries. Their German speaking ancestors emigrated in the 13th and 14th century from the upper Rhone valley to the Safian valley (Safiental) in the canton of Grison (Graubünden). This population remained quite isolated through the language, and by the geographic situation of the Safian valley. VIELI in 1855 (cited in DUCKERT & KOLLER 1975) described the disease to be exclusively transmitted by women who were called "Conductoren" (which means carriers according to the local expression), whereas affected men were called "Bluter" in German (bleeders). However, it later turned out that this most intensively investigated family with its long recognized mode of inheritance - via carriers - suffered not from the classical hemophilia A, but rather from the rarer hemophilia B.

The bleeder family of Wald (canton of Zürich) is a good example of hemophilia A. This family, unfortunately, was less well investigated, its description being found in the thesis of HANS STAHEL from 1880 (cited in DUCKERT & KOLLER 1975).

Genetics

Although the X-chromosome related defect of hemophilia A has been known for a long time, the complete characterization and cloning of F VIII was not reported before 1984 (GITSCHIER et al. 1984, WOOD et al. 1984, VEHAR et al. 1984, TOOLE et al. 1984). The molecular background of hemophilia has been extensively examined and the related gene defects have been well established. Single nucleotide changes have recently been reported as causes of hemophilia (GITSCHIER et al. 1985, ANTONARAKIS et al. 1985, YOUSSOUFIAN et al. 1986). These point mutations leading to an abnormal stop codon occur in all regions of the F VIII gene, thus preventing the transcription of the entire gene. The affected patients suffer from a severe hemophilia. In at least one case, however, a point mutation resulted in an amino acid substitution and the hemophiliac had only a mild form of this disease (GITSCHIER et al. 1986). Deletions (GITSCHIER et al. 1985, YOUSSOUFIAN et al. 1987) and, more recently, insertions of long, interspersed repetitive elements (L1) in the F VIII gene have also been found to be causes of hemophilia (KAZAZIAN et al. 1988).

Patients who develop inhibitors to the coagulaton factors VIII or IX (for an overview of "inhibitor patients" see HOYER 1984), suffer from a bleeding tendency of variable severity. A genetic predisposition to inhibitor formation, if existent, is not yet fully understood (SHAPIRO 1984).

Health care and social problems

The life of a hemophiliac is accompanied by many difficulties. With the increased mobility of the child, hemorraghes occur more frequently. Lengthy stays in hospitals, even for minor surgery, like tooth extraction or after minor injuries, raise the costs of the medical care enormously. Because of frequent absences hemophilic children often have problems at school and their social integration is difficult. Particular patient care, including medical support, is therefore an important medical aspect.

Therapy

Replacement therapy with blood-derived F VIII concentrates is, so far, the only way to manage severe hemophilia. The rapid development of new plasma fractionation techniques, such as chromatography or immunoabsorbance, has resulted in highly purified factor concentrates. The yields of extraction have been considerably improved during the past several years. The risks of viral contaminations such as hepatitis, non A / non B hepatitis and AIDS, however, have also increased. The safety of blood-derived concentrates was the most discussed topic during the symposium of transfusion medicine, held in Zürich in September 1988 (MORGENTHALER 1988, STEPHAN 1988, SUGG 1988). All available techniques for decontamination of plasma products display the same disadvantages: loss of yield and protein activity of up to 50%.

It seems likely that the isolation of at least some blood clotting factors and other plasma fractions from human plasma will be possible through the use of recently developed genetic and cell engineering methods (LAWN & VEHAR 1986, KINGSMAN & KINGSMAN 1988). Despite this glimmer of hope to obtain safe and inexpensive plasma concentrates in the future, it cannot be guaranteed that the rising demands for these products will be met worldwide.

Vasoactive substances of various generic groups such as catecholamines (INGRAM 1961, INGRAM & VAUGHAN-JONES 1966), vasopressins and nicotinic acid (CASH et al. 1974, MANNUCCI et al. 1975, PROWSE et al. 1979) elicit an increase of blood clotting F VIII in man. The treatment with deamino-D-arginine-vasopressin (DDAVP), an analogue of the neurohypophyseal hormone vasopressin (ZAORAL et al. 1967) has already been well established in mild or moderate hemophilia and in von Willebrand's disease (MANNUCCI et al. 1977a, MANNUCCI et al. 1977b, KOBAYASHI et al. 1978). This peptide was originally designed to substitute for vasopressin in diabetes insipidus (PLIŠKA 1985), and its hemostatic effect was detected accidentally some ten years ago. The mechanism of F VIII and F IX enhancement remains unexplained. A direct effect on *de novo* synthesis, or liberation of stored coagulation factors, can be excluded (MOFFAT et al. 1984). F VIII activities after DDAVP administration are often high enough to prevent hemorrhagic incidents during or after small surgery (MANNUCCI et al. 1977a, SCHIMPF & ROTHMANN 1980). DDAVP has also been recently found to shorten the prolonged skin bleeding times that occur with uremia, cirrhosis, and platelet dysfunction of various causes. Even in hemostatically normal persons undergoing spinal fusion surgery or cardiopulmonary bypass surgery, DDAVP reduces blood loss and the amount of blood required for transfusion (for review, see MANNUCCI 1988). Finally, the administration of DDAVP to blood donors to enhance blood coagulation yield, has been discussed (NILSSON et al. 1979, VILHARDT & NILSSON 1988).

However easy the application of DDAVP might be (intranasal self-administration is possible), this analog is not an ideal drug for these purposes because of its numerous side effects. These include antidiuresis for several hours and, as a consequence, change in plasma osmolality after infusion (SEILER & HERTLEIN 1980), facial flushing, hypotension, tachycardia (BICHET et al. 1988) and psychogenic stress (BLÄTTLER 1980). Peptides with even greater antihemophilic effects and fewer side effects would be desirable.

Hemophilia in animals

Domestic animals

A detailed review of bleeding disorders in other mammalian species was published recently (GRÜN 1985). Hemophilia A in dogs has been described in very different purebreeds as well as in crossbreeds (FIELD et al. 1946, BROCK et al. 1963, KANEKO et al. 1967, GILES et al. 1982). Hemophilic cats (COTTER et al. 1978), horses (ARCHER 1961, SANGER et al. 1976, HUTCHINS et al. 1967) and cattle (HEALY et al. 1983) are known. Hemophilic animals suffer from the same symptoms as their human counterparts, and the inheritance pattern of the disease is also recessive and sex-linked (X-linked).

Recent findings with sheep

In the flock of White Alpine sheep at our experimental station Chamau (canton of Zug), several male offspring of one ewe (No. 157; Fig. 3) were born with a bleeding disorder (PLIŠKA et al. 1982). The first of these lambs died within a few hours *post partum*, due to bleeding from the umbilical cord. Daughters (No. 665, No. 711) and grand-daughters (No. 1587, No. 1776) of the same ewe gave birth to other male lambs with a bleeding disorder. Lambs which survived longer suffered from either massive hematomas (primarily in the muscles of their hind legs) or from spontaneous hemarthroses (especially in weight-bearing joints). As a consequence of these occurrences these animals showed great pain in standing up for suckling and walking (Fig. 4).

The prolonged activated prothrombin time (APPT, a global coagulation test) of the affected lambs indicated a disorder of the intrinsic pathway of the coagulation cascade. Further examination of the intrinsic pathway showed that the pathologic APTT was due to the very low F VIII activity of about 2% of the norm, thus indicating a severe hemophilia A. Other hematological findings such as number of platelets and activity of the coagulation factors V, VII, IX, and X, remained within the physiological range.

Transfusion of normal plasma or the application of cryoprecipitated antihemophilic globulins from sheep plasma (prepared by us) diminished the hemorrhages for a certain time, but none of the lambs exceeded the age of two months. In all cases death occurred within a few hours after a severe bleeding incident. The heterozygous carriers show a considerable scatter of F VIII activity. Carrier detection by methods of multivariate analysis of clotting data, therefore, carries a high risk of misclassification. An important reason for this variability in carriers may be the accidental inactivation (Lyonization) of one X chromosome in an early stage of embryogenesis (GRAHAM et al. 1975).

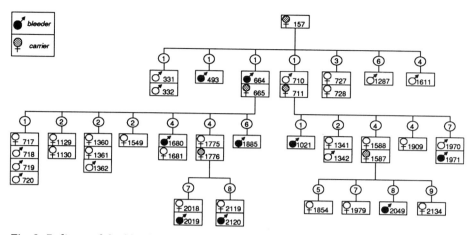

Fig. 3. Pedigree of the bleeder flock of White Alpine sheep. Numbers in circles indicate the sires.

Use of animal models

Spontaneous animal models of human hemophilia can be employed for various purposes. Some of those are mentioned below.

Genetic studies

The lack of manifest hemophilia in the female has been the subject of speculation for several decades. The available animal model of hemophilia in dogs (BRINKHOUSE & GRAHAM 1950) demonstrated that the breeding of hemophilic female dogs is possible, thus supplying the proof that the homozygote appearance of the mutated gene is not lethal. The female bleeders resemble in pathological terms, the affected males, and the theoretical 50% occurrence of bleeder females in the offspring population was demonstrated.

It would indeed be important to know the differences between human and animal coagulation factors at the molecular level. The application of several breeding strategies to an animal model might detect further relatonships between the genetic background and the mechanisms of blood clotting regulation.

OHNO (1969) postulated an extremely high conservation of X-linked genes in mammals. By using cDNA probes of homologue X-linked sequences in different species (e.g. F VIII and others), it is possible to construct a genetic map of the X chromosome in order to determine conserved homologous linkages and to identify sites of rearrangement (MULLINS et al. 1988).

Therapy with recombinant F VIII

Previous studies have clearly demonstrated the in vitro clotting activity of cloned F VIII (WOOD et al. 1984, EATON et al. 1986, EATON et al. 1987). GILES and co-workers (1988b) provided evidence that recombinant F VIII possesses full functional activity in the hemophilic dog and exhibits normal recovery and survival characteristics. Because of the extremely low F VIII activity in hemophilic sheep, these animals would be an even better animal model to test recombinant F VIII activity *in vivo*. The high mortality rate has, however, until now prevented the successful employment of male bleeders.

F VIII inhibitor studies

Approximately 10-15 % of all patients with hemophilia A develop antibodies to F VIII after a period of replacement therapy. In this situation therapeutic alternatives are required. Several authors (KINGDON & HASSELL 1981, O'BRIEN et al. 1988, GILES et al. 1988a) have presented the hemophilic

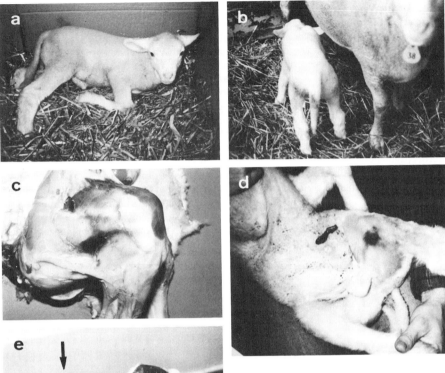

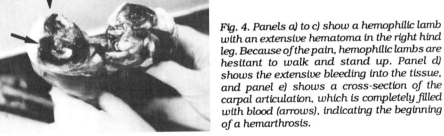

Fig. 4. Panels a) to c) show a hemophilic lamb with an extensive hematoma in the right hind leg. Because of the pain, hemophilic lambs are hesitant to walk and stand up. Panel d) shows the extensive bleeding into the tissue, and panel e) shows a cross-section of the carpal articulation, which is completely filled with blood (arrows), indicating the beginning of a hemarthrosis.

dog as an animal model to evaluate the hemostatic effectiveness of materials designed to bypass F VIII inhibitors. The treatment of "inhibitor patients" with such a material carries high risks of thrombosis, because it contains activated coagulation factors, or tissue factors. Even more serious is the risk of anticoagulant activity of the same material by the activation of protein C, which inactivates primarily F V and F VIII *in vivo* (cf. ref. above), leading to severe bleeding disorders. These narrow limits of therapeutic effectiveness indicate the need for testing any new preparations in an animal model. Again, hemophilic sheep might serve these purposes well.

Screening of antihemophilic drugs

Investigations of new antihemophilic drugs require suitable animal models and usually involve substantial difficulties. First, the blood withdrawal must be done under optimal conditions, without extensive injury to the punctured vein and avoiding any stress to the animal, including anesthesia etc. These conditions exclude the use of small laboratory animals. Second, species differences in blood clotting, concentration of clotting factors, and the number and function of platelets may pose additional problems. Third, drug effects upon the increase of F VIII plasma levels may considerably differ in normal and hemophilic individuals. Therefore, even a "normal" human may not be a satisfactory model. Finally, the mechanisms by which the vasoactive substances enhance blood clotting factor VIII still need to be elucidated.

We have attempted to employ normal sheep as a possible animal model for the screening of antihemophilic drugs (HEINIGER et al. 1988). Advantages of the sheep as an animal model are obvious. The occurrence of hemophilia A with a clinical pathology and inheritance, comparable to that of the human disease, suggests a high degree of analogy between both coagulation systems. Handling of the animal is also simple, a series of blood samples can be taken without adversely affecting the animal's health and, during a pharmacological investigation, the animals can stay together in small groups in their habitual environment, thus minimalizing external stressors.

The principal difference in the hemostatic system of animals, compared to that of humans, is the level of coagulation factors (HAWKEY 1975). F VIII in sheep, for instance, displays a four to five times higher activity. The antihemophilic response to an investigated drug, measured in relation to basal F VIII level, is correspondingly smaller in sheep. Employment of female carriers in the experiments with roughly 50% of F VIII concentrations, may partially compensate for this drawback.

There exists, however, a subpopulation of animals within the employed flock which does not display a regular antihemophilic response. These animals can be detected by a preliminary test with DDAVP, and ought to be excluded from further experiments.

Conclusions

We have observed that hemophilia A (F VIII deficiency) occurs in sheep of the White Alpine race (WAS). The inheritance pattern, the hematologic findings and the pathologic alterations in sheep hemophilia are consistent with those observed in man. The affected bleeder animals could serve as a model to test recombinant coagulation factor VIII *in vivo* and to reveal the molecular mechanisms of the defect. Moreover, after five years of experience we consider normal sheep or female carriers to be suitable animal models for the screening of antihemophilic drugs which a) are important alternatives in the treatment of various bleeding disorders, or b) which may help to reduce blood loss and the transfusion requirements of hemostatically normal persons during surgery.

Acknowledgements

The work was supported by a research grant of the Swiss Federal Institute of Technology, ETH Zürich, and of the Sandoz Foundation, Basel.

References

ANTONARAKIS, S. E.; WABER, P. G.; KITTUR, S. D.; PATEL, A. S.; KAZAZIAN, H. H., Jr.; MELLIS, M. A.; COUNTS, R. B.; STAMATOYANNOPOULOS, G.; BOWIE, E. J. W.; FASS, D. N.; PITTMAN, D. D.; WOZNEY, J. M.; TOOLE, J. J. (1985): Hemophilia A: detection of molecular defects and of carriers by DNA analysis. N. Engl. J. Med. 313, 842-848.

ARCHER, R. K. (1961): True haemophilia (haemophilia A) in a thoroughbred foal. Vet. Rec. 73, 338-340.

BICHET, D.; RAZI, M.; LONERGAN, M.; ARTHUS, M. F. (1988): Factor VIIIc (F VIIIc), von Willebrand factor (vWF), mean arterial blood pressure (MAP), pulse, and renin (PRA) responses to dDAVP infusion in patients (pts) with congenital nephrogenic diabetes insipidus (CNDI). Kidney Int. 33, 182 (Abstract).

BLÄTTLER, W. (1980): Psychic effects of 1-deamino-8-arginine vasopressin (DDAVP). In: MINIRIN DDAVP-Anwendung bei Blutern (Editor: SUTOR, A. H.). F. K. Schattauer Verlag, Stuttgart, N.Y.. P. 173.

BRINKHOUSE, K. M.; GRAHAM, J. B. (1950): Hemophilia in the female dog. Science 111, 723-724.

BROCK, W. E.; BUCKNER, R. G.; HAMPTON, J. W.; BIRD, R. M.; WULZ C. E. (1963): Canine hemophilia. Arch. Pathol. 76, 464-469.

CASH, J. D.; GADER, A. M. A.; DA COSTA, J. (1974): The release of plasminogen activator and factor VIII by LVP, DDAVP, AT III and OT in man. Brit. J. Haematol. 27, 363-364.

COTTER, S. M.; BRENNER, R. M.; DODDS, W. J. (1978): Hemophilia A in three unrelated cats. J. Amer. Vet. Med. Assoc. 172, 166-168.

DAVIE, E. W.; RATNOFF, O. D. (1964): Waterfall sequence for intrinsic blood clotting. Science 145, 1310-1312.

DUCKERT, F.; KOLLER, F. (1975): The old swiss haemophilia families of Tenna and Wald. In: Handbook of Hemophilia, Part 1 (Editors: BRINKHOUSE, K. M.; HEMKER, H. C.). Excerpta Medica, Amsterdam. American Elsevier Publishing Company Inc., N.Y.. Pp. 21-29.

EATON, D. L.; WOOD, W. I.; EATON, D.; HASS, P. E.; HOLLINGSHEAD, P.; WION, K.; MATHER, J.; LAWN, R. M.; VEHAR, G. A.; GORMAN, C. (1986): Construction and characterization of an active factor VIII variant lacking the central one-third of the molecule. Biochemistry 25, 8343-8347.

EATON, D. L.; HASS, P. E.; RIDDLE, L.; MATHER, J.; WIEBE, M.; GREGORY, T.;VEHAR, G. A. (1987): Characterization of recombinant human factor VIII. J. Biol. Chem. 262, 3285-3290.

FIELD, R. A.; RICKARD, C. G.; HUTT, F. B. (1946): Hemophilia in a family of dogs. Cornell Vet. 36, 285-300.

GILES, A. R.; MANN, K. G.; NESHEIM, M. E. (1988a): A combination of factor Xa and phosphatidylcholine-phosphatidylserine vesicles bypasses factor VIII *in vivo*. Brit. J. Haematol. 69, 491-497

GILES, A. R.; TINLIN, S.; GREENWOOD, R. (1982): A canine model of hemophilic (factor VIII: C deficiency) bleeding. Blood 60, 727-730.

GILES, A. R.; TINLIN, S.; HOOGENDOORN, H.; FOURNEL, M. A.; NG, P.; PANCHAM, N. (1988b): *In vivo* characterization of recombinant factor VIII in a canine model of hemophilia A (factor VIII deficiency). Blood 72, 335-339.

GITSCHIER, J.; WOOD, W. I.; GORALKA, T. M.; WION, K. L.; CHEN, E. Y.; EATON, D. H.; VEHAR, G. A.; CAPON, D. J.; LAWN, R. M. (1984): Characterization of the human factor VIII gene. Nature 312, 326-330.

GITSCHIER, J.; WOOD, W. I.; SHUMAN, M. A.; LAWN, R. M. (1986): Identification of a missense mutation in the factor VIII gene of a mild hemophiliac. Science 232, 1415-1416.

GITSCHIER, J.; WOOD, W. I.; TUDDENHAM, E. G. D.; SHUMAN, M. A.; GORALKA, T. M.; CHEN, E. Y.; LAWN, R. M. (1985): Detection and sequence of mutations in factor VIII gene of haemophiliacs. Nature 315, 427-430.

GRAHAM,J. B.; BARROW, E. S.; ELSTON, R. C. (1975): Discussion paper: Lyonization in hemophilia: a cause of error in direct detection of heterozygous carriers. In: Recent Advances in Hemophilia (Editor: ALEDORT, L. M.). Ann. NY. Acad. Sci. 240, 141-146.

GRÜN, E. (1985): Erbliche Blutgerinnugsdefekte bei Haustieren. Mh. Vet. Med. 40, 746-749.

HAWKEY, C. M. (Editor) (1975): Comparative Mammalian Haematology. Cellular Components and blood coagulation of captive wild animals. Heinemann Medical Books LTD, London.

HEALY, P. J.; SEWELL, C. A.; EXNER, T.; MORTON, A. G.; ADAMS, B. S. (1984): Haemophilia in Hereford cattle: factor VIII deficiency. Aust. Vet. J. 61, 132-133.

HEINIGER, J.; KISSLING-ALBRECHT, L.; NEUENSCHWANDER, S.; RÖSLI, R.; PLIŠKA, V. (1988): Antihaemophilic effect of vasopressin, deamino-(D-arginine^8)-vasopressin and adrenaline in sheep: Proposal for an *in vivo* assay system. Brit. J. Pharmacol. 94, 279-281.

HOYER, L. W. (Editor) (1984): Factor VIII Inhibitors. A. R. Liss, Inc., N.Y., Prog. Clin. Biol. Res. 150.

HUTCHINS, D. R.; LEPHERD, E. E.; CROOK, I. G. (1967): A case of equine haemophilia. Aust. Vet. J. 43, 83-87.

INGRAM, G. I. C (1961): Increase in antihaemophilic globulin activity following infusion of adrenaline. J. Physiol. 156, 217-224.

INGRAM, G. I. C.; VAUGHAN-JONES, R. (1966): The rise in clotting factor VIII induced in man by adrenaline: Effect of α- and ß-blockers. J. Physiol. 187, 447-454.

KANEKO, J. J.; CORDY, D. R.; CARLSON, G. (1967): Canine hemophilia resembling classic hemophilia A. J. Amer. Vet. Med. Assoc. 150, 15-21.

KAZAZIAN, H. H., Jr.; WONG, C.; YOUSSOUFIAN, H.; SCOTT, A. F.; PHILIPS, D. G.; ANTONORAKIS, S. E. (1988): Haemophilia A resulting from *de novo* insertion of L1 sequences represents a novel mechanism for mutation in man. Nature 332, 164-166.

KINGDON, H. S.; HASSELL, T. M. (1981): Hemophilic dog model for evaluating therapeutic effectiveness of plasma protein fractions. Blood 58, 868-872.

KINGSMAN, S. M.; KINGSMAN, A. J. (Editors) (1988): Genetic Engineering. An introduction to gene analysis and exploitation in eukaryotes. Blackwell Scientific Publications, Oxford, London, Edinburgh, Boston, Palo Alto, Melbourne. Pp. 431-436.

KOBAYASHI, I.; ITO M.; SHIBATA, A.; SHINADA, S.; TSUKADA, T. (1978): D.D.A.V.P. in haemophilia B. Lancet 1, 615-615.

LAWN, R. M.; VEHAR, G. A. (1986): The molecular genetics of hemophilia. Sci. Amer. 254 (3), 40-46.

MACFARLANE, R. G. (1964): Haematology: an enzyme-cascade in the blood clotting mechanism and function as a biochemical amplifier. Nature 202, 498-499.

MANNUCCI, P. M. (1988): Desmopressin: a non transfusional form of treatment for congenital and acquired bleeding disorders. Blood 72, 1449-1455.

MANNUCCI, P. M.; ÅBERG, M.; NILSSON I. M.; ROBERTSON, B. (1975): Mechanism of plasminogen activator and factor VIII increase after vasoactive drugs. Brit. J. Haematol. 30, 81-93.

MANNUCCI, P. M.; RUGGERI, Z. M.; PARETI, F. I.; CAPITANIO, A. (1977a): 1-deamino-8-D-arginine vasopressin: A new pharmacological approach to the management of haemophilia and von Willebrand's disease. Lancet 1, 869-872.

Mannucci, P. M.; Ruggeri, Z. M.; Pareti, F. I.; Capitanio, A. (1977b): D.D.A.V.P. in Haemophilia. Lancet 2, 1171-1172.

Moffat, E. H.; Giddings, J. C.; Bloom, A. L. (1984): The effect of desamino-D-arginine vasopressin (DDAVP) and naloxone infusions on factor VIII and possible endothelial cell (EC) related activities. Brit. J. Haematol. 57, 651-662.

Morgenthaler, J. J. (1988): Virusinaktivierung und Virussicherheit stabiler Plasmaprodukte. In: Deutsche Gesellschaft für Transfusionsmedizin und Immunhämatologie Zürich, Symposium, 14.-18. September 1988. P. 24 (Abstract).

Mullins, L. J.; Grant, S. G.; Stephenson, D. A.; Chapman, V. M. (1988): Multilocus molecular mapping of the mouse X chromosome. Genomics 3, 187-194.

Nilsson, I. M.; Walter, H.; Vilhardt, H. (1979): Factor VIII concentrate prepared from DDAVP stimulated blood donor plasma. Scand. J. Haematol. 22, 42-46.

O'Brien, D. P.; Giles, A. L.; Tate K. M.; Vehar, G. A. (1988): Factor VIII-bypassing activity of bovine tissue factor using the canine hemophilic model. J. Clin. Invest. 82, 206-211.

Ohno, S. (1969): Evolution of sex chromosomes in mammals. Ann. Rev. Genet. 3, 495-524.

Pliška, V.; Schwander, B.; Allmendinger, W.; Müller-Lhotsky, A. (1982): Erbliche Hämophilie beim Schaf: Untersuchungen an potentiellen Konduktorinnen. Schweiz. Landw. Monatshefte 60, 284-295.

Pliška, V. (1985): Pharmacology of deamino-D-arginine vasopressin. In: Diabetes Insipidus in Man. (Editors: Czernichow, P.; Robinson, A. G.). Front. Hormone Res. 13, 278-291.

Prowse, C. V.; Sas, G.; Gader, A. M. A.; Cort, J. H.; Cash, J. D. (1979): Specificity in the factor VIII response to vasopressin infusion in man. Brit. J. Haematol. 41, 437-447.

Rosner, F. (1969): Hemophilia in the Talmud and rabbinic writings. Ann. Intern. Med. 70, 833-837.

Sanger, V. L.; Mairs, R. E.; Trapp, A. L. (1964): Hemophilia in a foal. J. Amer. Vet. Med. Assoc. 144, 259-264.

Schimpf, K.; Rothmann, P. (1980): Behandlung mit DDAVP während 18 Zahnextraktionen, 2 Muskelblutungen, 1 Bisswunde und einer Kiefernhöhlenspülung bei insgesamt 8 Patienten mit milder Hämophilie A. In: MINIRIN, DDAVP-Anwendung bei Blutern (Editor: Sutor, A. H.). F. K. Schattauer Verlag Stuttgart, N.Y.. Pp.131-136.

Seiler, G.; Hertlein, W. (1980): Kontrolle von Osmolarität, Ionogramm und Serum-Protein-Spiegel nach intravenöser und nasaler Anwendung von DDAVP. In: MINIRIN, DDAVP-Anwendung bei Blutern. (Editor: Sutor, A. H.). F. K. Schattauer Verlag Stuttgart, N.Y.. Pp. 160-166.

Shapiro, S. S. (1984): Genetic predisposition to inhibitor formation. In: Factor VIII Inhibitors. (Editor: Hoyer, L. W.). A. R. Liss Inc., N.Y., Prog. Clin. Biol. Res. 150, 45-55.

Stephan, W. (1988): Probleme bei der Sterilisation labiler Plasmaproteine. In: Deutsche Gesellschaft für Transfusionsmedizin und Immunhämatologie Zürich, Symposium, 14.-18. September 1988. P. 25 (Abstract).

Sugg, U. (1988): Virussicherheit von frisch gefrorenem Plasma. Deutsche Gesellschaft für Transfusionsmedizin und Immunhämatologie, Zürich, Symposium, 14.-18. September 1988. P. 26 (Abstract).

Toole, J. J.; Knopf, J. L.; Wozney, J. M.; Sultzman, L. A.; Buecker, J. L.; Pittman, D. D.; Kaufman, R. J.; Brown, E.; Shoemaker, C.; Orr, E. C.; Amphlett, G. W.; Foster, W. B.; Coe, M. L.; Knutson, G. J.; Fass, D. N.; Hewick, R. M. (1984): Molecular cloning of a cDNA encoding human antihaemophilic factor. Nature 312, 342-347.

Vehar, G. A.; Keyt, B.; Eaton, D.; Rodriguez, H.; O'Brian, D. P.; Rotblat, F.; Oppermann, H.; Keck, R.; Wood, W. I.; Harkins, R. N.; Tuddenham, E. G. D.; Lawn, R. M.; Capon D. J. (1984): Structure of human factor VIII. Nature 312, 337-342.

Vilhardt, H.; Nilsson, I. M. (1988): The effect of vasopressin analogues on factor VIII and plasminogen activator. In: Recent Progress in Posterior Pituitary Hormones (Editors: Yoshida, S.; Share, L.). International Congress Series, Excerpta Medica, Amsterdam. Pp. 177-184

Wood, W. I.; Capon, D. J.; Simonsen, C. C.; Eaton, D. L.; Gitschier, J.; Keyt, B.; Seeburg, P. H.; Smith, D. H.; Hollingshead, P.; Wion, K. L.; Delwart, E.; Tuddenham, E. G. D.; Vehar, G. A.; Lawn, R. M. (1984): Expression of active human factor VIII from recombinant DNA clones. Nature 312, 330-337.

Youssoufian, H.; Kazazian, H. H., Jr.; Phillips D. G.; Aronis, S.; Tsiftis, G.; Brown, V. A.; Antonarakis, S. E. (1986): Recurrent mutations in haemophilia A give evidence for CpG mutation hotspots. Nature 324, 380-382.

Youssoufian, H.; Antonarakis, S. E.; Aronis, S.; Tsiftis, G.; Phillips, D. G.; Kazazian, H. H., Jr. (1987): Characterization of five partial deletions of the factor VIII gene. Proc. Natl. Acad. Sci. USA. 84, 3772-3776.

Zaoral, M.; Kolc, J.; Šorm, F. (1967): Amino acids and peptides.LXXI. Synthesis of 1-deamino-8-D-aminobutyrine vasopressin, 1-deamino-8-D-lysine vasopressin and 1-deamino-8-D-arginine vasopressin. Coll. Czech. Chem. Commun. 32, 1250-1257.

Authors' address:
S. Neuenschwander and V. Pliška, Institut für Nutztierwissenschaften, ETH Zürich, CH-8092 Zürich, Switzerland.

Maintainance of anomalous animal lines

Reproductive biotechnology for preserving genetic anomalies in farm animals

U. Süss

Introduction

Various anomalies observed in farm animals, representing interesting research models, have been described in the previous chapters. The preservation of such animal models is not only desirable but also essential in order to perform extended research on the cause of the diseases or disorders. Often there is no analogous model available in laboratory animals.

Farm animals with extraordinary features of any kind are of great interest for basic research in genetics, molecular biology and physiology. Efforts to preserve any kind of rare farm and wild animals are undertaken by several organisations, for example: Food and Agriculture Organization of the United Nations (FAO), Rome, Italy; United Nations Environment Programm (UNEP), Nairobi, Kenya; Zentralstelle der Europäischen Vereinigung für Tierzucht, Hannover, G.F.R. The choice of the conservation method to be employed requires careful consideration due to the limited availability of the material of interest.

This chapter will therefore, review the methods of preservation applicable to farm animals, and describe the principle procedures used. The success rates to be expected at the present time will be mentioned, although significant improvements may arrive in the future for most of these techniques.The following techniques will be described: 1) animal breeding programmes; 2) artificial insemination; 3) embryo transfer; 4) *in vitro* maturation and fertilization of oocytes; 5) DNA-conservation.

Animal breeding programmes

Animal breeding programmes represent the oldest and most widely applied method to preserve and improve domestic, as well as wild animals. The breeding of farm animals aims in general at the improvement of particular traits such as yield and quality of milk or meat. Selective breeding sometimes results in unexpected or abnormal features of an individual animal, representing interesting research material. For research purposes animal models need to be preserved in their momentary state, with as little change as possible. Hence, prolonged breeding would not be beneficial, but is in some cases the only choice. The occurring genetic drift may then be limited by the application of a breeding programme with genetic anomalies . Such breeding programmes have been described for laboratory animals (Van Zutphen et al. 1988). It requires well defined and reliably measurable genetic markers as for example enzyme variants, blood groups, chromosome markers (i.e. C-band polymorphisms), molecular genetic marker (fingerprints, restriction fragment length polymorphisms). For a given programme, a set of such markers is selected and a standard value is determined. On each animal these markers are measured. The animals carrying markers which deviate the least from the standard are selected to be bred, thus producing the next generation with as little deviation from the standard as possible.

Extensive studies of the properties of a large population are a prerequisite to determine the markers and their standard values. Great efforts are at present undertaken to find more genetic markers in farm animals to establish well defined gene maps, also for these species.

Animal breeding programmes serve well to replace and supply live animals continuously. However, these programmes do not preserve the genome unchanged over time.

Artificial insemination

The introduction of artificial insemination (AI) revolutionized the breeding of farm animals. The achievement of successful freezing and thawing of semen allowed the storage of one half of the genome over years. Artificial insemination is routinely performed with frozen/thawed semen in the cattle. In the pig and the horse artificial insemination with fresh semen is more successful. The success rate of AI is expressed as the non-return rate which is the percent of females not returning to oestrus and therefore not requiring another insemination within 60 to 90 days. Non-return rates for normal bulls are between 65 - 70%. The semen to be employed in AI programmes is collected in an artificial vagina from selected males. The ejaculate is carefully examined for volume and colour, sperm count, motility and morphology of spermatozoa. After freezing and thawing, sperm motility and morphology are assessed again (FOOTE 1974, GARNER 1984). In addition, the performance of the offspring is recorded to prevent the spreading of unwanted traits or defects. The history of each animal can easily be traced back in the herd book.

The procedure for the freezing of semen includes the dilution of the ejaculate with a cryoprotecting diluent, which differs for each species, and placement into straws holding variable amounts of diluted semen. The sperm count must be adjusted to the needs of the fertilization optimum and results in a large number of straws for each ejaculate and male (PAUFLER 1974).

For insemination the straw containing frozen/thawed semen is placed into a catheter designed for this purpose. The catheter is introduced through the vagina and in some species through the cervix to reach into the uterus cavity, where the semen is deposited. A catheter is readily passed through the cervix of cows and pigs, whereas the sheep and goat cervices are impenetrable due to their anatomy. The cervical canal of the sheep is bent and several folds of cervical tissue further obstruct the introduction of a catheter (Fig. 1). For these species AI is not very successful since the semen has to be placed at the entrance of the cervix, which results in higher losses of spermatozoa and lower pregnancy rates. Sheep and goat semen may be frozen and thawed successfully, still representing a helpful means of preserving interesting genetic information.

Fig. 1. The anatomy of the cervical canal of a sheep. Note the angle of about 90 degree (arrow), hindering the introduction of a catheter. (Cast preparation by D. Stauffer, and drawn by P. Brouwer, Institute of Animal Science, ETH-Zürich).

Although the storage of frozen semen is very valuable and quite easy to achieve, the female to be inseminated must be selected carefully in order not to interfere with the features of interest. Since the genome of the inseminated female has not been preserved, further breeding will be necessary if homozygous individuals are required.

Embryo transfer

The technology of embryo transfer including the freezing of embryos for storage has been established mainly to improve animal breeding programmes. It also offers ways to preserve embryos containing extraordinary genetic information, or embryos of rare and endangered species. Embryos recovered during the transfer procedure represent a valuable research material for studies of embryo development in general and genetic aspects in particular. By freezing embryos, the complete genetic information derived from both parents is preserved, in contrast to the storage of spermatozoa. This may be of importance depending on the subject under study.

The embryo transfer procedure includes several steps and techniques, each requiring optimal performance to guarantee good results. Since embryo transfer is employed most frequently in cattle, this chapter describes the techniques used for this species. The steps are: superovulation of the donor animal, embryo recovery, embryo qualification, embryo transfer, synchronization of recipients, embryo freezing, embryo storage and embryo thawing.

Superovulation

The collection of several embryos from one donor cow is achieved by the application of a superovulating treatment. Various regimens for the administration of the stimulating hormones, such as follicle stimulating hormone (FSH) or pregnant mare serum gonadotropin (PMSG) have been described (WISE et al. 1986, SEIDEL 1981). The hormones induce the maturation of several follicles during the stimulated cycle, resulting in the ovulation of multiple, fertilizable oocytes. Superovulation regimens also include a synchronization treatment. The injection of prostaglandin F$_{2\alpha}$ induces the breakdown of the corpus luteum, resulting in a surge of luteinizing hormone (LH) and subsequent ovulation within approximately 48 to 72 hours.

Table 1. *Response of individual cows to hormonal stimulation and synchronization.*

time of slaughter post PGF^a	no. of mature oocytes f/o^b	time of LH surge post PGF^a	time of oestrus post PGF^a	oocytes penetrated^c f/o^b
72	1 / 13	42	46.5	0 / 11
72	8 / 4	48	-	0 / 0
72	5 / -	-	31.5	2 / -

^{a} PGF = Prostaglandin F$_{2\alpha}$.

^{b} Number of oocytes recovered from follicles (f) and oviduct (o).

^{c} Number of oocytes penetrated by a single sperm.

The response to superovulation and synchronization of individual cows varies considerably. It is obvious from Table 1 that synchronization of the LH surge and the resulting onset of oestrous behaviour was not very precise. The numbers of mature oocytes recovered from Graafian follicles and from the oviduct differed greatly between animals. The result of *in vitro* fertilization of such oocytes expressed as the number of oocytes penetrated by a single sperm also varied considerably among individual animals. The presence of chromosomal mutations may be an additional hindrance to the production of fertilizable oocytes, which may or may not be overcome by superovulation treatments. The data presented in Table 1 are meant to demonstrate the difficulties encountered when individual animals are concerned. By contrast, the application of superovulation in herds with large animal numbers is very efficient and results in multiple calves (2 to 20) per donor cow within one breeding cycle (SEIDEL 1981).

Following superovulation, artificial insemination is employed (see previous section). The synchronization schedule, as well as observation of oestrous behaviour, determine the appropriate time for insemination.

Embryo recovery and qualification

Six to seven days after successful insemination the embryos can be flushed out of the uterus. For this purpose, a specially designed catheter is introduced into the uterus via the vagina. A small rubber bulb at the near end of the catheter is inflated to seal the uterus towards the cervix to prevent the flushing medium from leaking out through the cervix. The flushing solution is forced in, suspending the embryos and carrying them out with the medium through an opening in the catheter. At day six or seven, normally developed embryos have reached the stage of morulae or blastocysts. Not all of the recovered embryos are of high quality and have reached these stages, even unfertilized oocytes are found. Therefore, careful quality control of the recovered embryos is necessary (HARTMANN 1983).

Embryos may be characterized morphologically, cytogenetically and by various enzymatic tests. Embryos to be transferred are generally qualified by their morphological appearance. Morphological criteria of a good embryo are (ADAMS 1982): equal size of blastomeres; no excluded blastomeres; homogenous granulation of the cytoplasm and in the case of blastocysts, a well defined blastocoel.

For cytogenetic analysis, a small number of cells of a morula or a blastocyst is removed by micromanipulation and placed in culture medium. The cells are stimulated to divide and then blocked at the metaphase stage by colcemid (ROTTMANN 1987). Karyotypes, chromosomal mutations as well as in situ hybridizations and the allocation of individual genes to a chromosome may be studied on such material (CHURCH et al. 1985).

Transfer and synchronization of the recipients

The embryo to be transferred is placed into a minitube, which then is fixed within a catheter similar to the one used for insemination. The catheter is introduced into the uterus via the vagina and cervix and the embryo is placed high up into the uterine horn. The immediate transfer of the recovered embryos to synchronized recipients results in the highest pregnancy rate of 65 - 70 %, as determined at 60 days after transfer. The transfer of two embryos into one recipient assures a high success rate. No great problems arise from the resulting twin pregnancies. The development of a pregnancy is not only dependent on the quality of the embryo but also on the readiness of the uterine environment for implantation. Therefore, the estrus cycle of the recipient cow has to be synchronized within the cycle stage of the donor animal. An asynchronism of one day between the donor's and the recipient's estrus cycle results in pregnancy rates similar as in those animals which have been exactly synchronized (SEIDEL 1981). The recipient has not been stimulated nor inseminated, to ascertain that the calf born was derived from transfer and thus from the desired parents. Cows of lesser quality are selected as recipients to carry calves from superior donors.

The procedure for the transfer of treated embryos is similar to that described above for nontreated embryos. However, depending on the treatment, the survival potential of the embryo will probably be negatively affected as expressed by a reduced pregnancy rate (Fig. 2). Interestingly, the splitting of embryos into two equal half-embryos slightly reduced the survival rate of the half embryos, but the overall pregnancy rate of about 120 % justifies the use of this method to increase the number of calves born. None of the calves developed from half-embryos showed any malformations as a result of the splitting.

Freezing, storage, thawing

The preservation of embryos, with or without genetic anomalies, is achieved by freezing the embryos and storing them in liquid nitrogen (WILLADSEN 1977, LEIBO 1986).

To prevent excessive ice crystal formation and thereof resulting cell damage, the embryos are equilibrated in protective media containing glycerol or dimethylsulfoxide and cooled at a controlled rate. At -7°C the freezing of the embryo containing medium, usually within minitubes, is induced by touching the straw with a very cold metallic instrument. This procedure is called seeding. Further cooling can be done more rapidly. After reaching a temperature of less than -30°C, the embryos are plunged into liquid nitrogen, where they can be stored over several years. Various procedures and machines have been developed to simplify the freezing of embryos, even less controlled cooling rates can be employed with satisfying results.

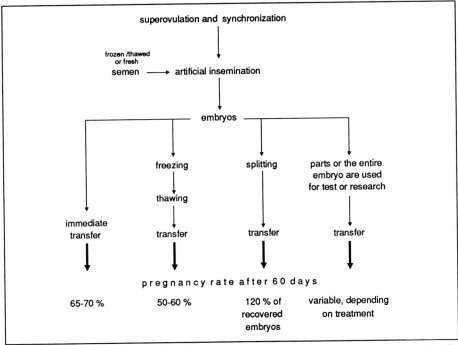

Fig. 2: Summary of possible procedures which can be applied to embryos recovered at the morula or blastocyst stage from donor cows or produced by in vitro fertilization procedures. (Pregnancy rates refer to results obtained with embryos recovered from donors).

Frozen embryos are reactivated by inserting the straw into a water bath at 20° C immediately upon removal from liquid nitrogen. The embryos are then transferred through a serial dilution of glycerol into fresh medium. At this point, quality control and other manipulations may be performed on the embryo before it is transferred to a synchronized recipient. The removal of the freeze-protectant is omitted in methods described more recently (LEIBO 1986, RALL & MEYER 1989). Here, the solution containing the embryo is mixed within the straw with a solution containing glucose which was also placed within the same straw, separated from the embryo by an air bubble. The embryo is then transferred to the recipient without examination, thus reducing the handling of the embryo considerably. This technique is called "one step method".

An average pregnancy rate of 50 to 60% is achieved with frozen, but otherwise untreated embryos.

In vitro maturation and fertilization

Access to early embryos and pronuclear stage ova is of great importance for gene manipulation as well as for studies on genetic control of development.

Hormonal stimulation, similar to the procedure described in the embryo transfer section, induces the growth of multiple ovulatory follicles. Shortly before the expected time of ovulation the mature but unfertilized oocytes are aspirated from the follicles, and already ovulated oocytes are flushed from the oviduct. Fertilized ova can be collected from the oviduct one day after insemination. Both procedures require hormonal treatment and surgery performed on the donor.

Another source of oocytes, although immature, is the ovaries of slaughtered female animals. Such ovaries still contain many oocytes at various developmental stages. Secondary follicles, which have formed an antrum, are visibly located at the surface and are readily aspirated to recover the oocytes. Throughout the cycle, follicles grow and degenerate, this degeneration process is called atresia. Follicles larger than approximately 6 mm in diameter may also be aspirated but their oocytes or granulosa cells often show signs of atresia. In the cow, only one or two tertiary follicles having reached a particular stage at a certain time of the ovarian cycle will develop to Graafian follicles and ovulate after the LH surge. Other large follicles will undergo atresia. In vivo, the oocyte of the Graafian follicle completes the first meiotic division upon stimulation by the LH surge. It ovulates at the metaphase II stage.

Oocytes recovered from small follicles are immature. For *in vitro* maturation the oocytes have to be evaluated and selected. The cumulus mass must be complete and compact and the cytoplasm should appear dark and evenly granulated, if recognizable through the cumulus cells (Fig. 3). Meiotic as well

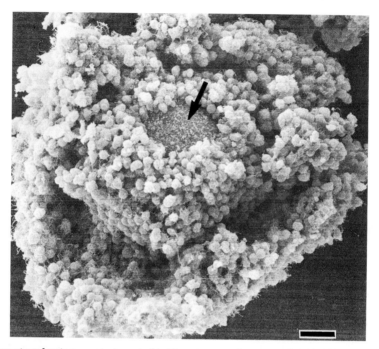

Fig. 3. *Immature bovine oocyte invested with a compact but incomplete cumulus mass. The surface of the zona pellucida is visible where the cumulus cells are missing (arrow). (Scanning electron micrograph taken by J. Kassner, Institute of Cell Biology, ETH-Zürich). Bar = 25 µm.*

as cytoplasmic maturation have to be induced and supported by appropriate culture conditions before *in vitro* fertilization can be attempted. Although there are several reports about offspring born from *in vitro* fertilized cow, pig and sheep oocytes, the numbers and success rates are still low (BRACKETT et al. 1982, CHENG et al. 1986, FIRST & PARRISH 1987, XU et al. 1987).

In spite of the rather discouraging results, *in vitro* maturation and fertilization may be a last possibility to produce offsprings, or at least ova and early embryos, from a particular female.

Following, the principle procedures of *in vitro* maturation and fertilization are listed for the cow (BRACKETT 1983, BALL et al. 1983, LEIBFRIED-RUTLEDGE et al. 1987).

The crucial steps of the procedure which lead to a fertilized pronuclear stage ovum are: selection of healthy cumulus oocyte complexes; completion of the first meiotic division with extrusion of the first polar body; during the same time period, occurance of cumulus expansion and the morphologically not visible cytoplasmic maturation; pretreatment of the spermatozoa to induce capacitation; acrosome reaction and penetration of the spermatozoon into the oocyte; the fertilized ovum undergoes the second meiotic division, followed by the formation of the female pronucleus; at the same time sperm head decondensation and formation of the male pronucleus occurs.

Meiotic maturation, meaning completion of the first meiotic division, is achieved easily and can be followed by light microscopy. Time sequence and chromosome configurations are well known (Süss et al. 1988).

For oocytes carrying chromosomal mutations this may be a critical event which could result in incorrect separation of the homologous chromosomes and subsequently in oocytes not capable of fertilization.

By contrast, cytoplasmic maturation cannot easily be visualized by light microscopy and is not yet fully understood. It occurs during the same time period as cumulus expansion and meiotic maturation, but it is unclear whether these events are independent or not. Cytoplasmic maturation is a prerequisite for successful male pronucleus formation and embryonic development (EPPIG & WARD-BAILEY 1982, BRACKETT 1983, FIRST & PARRISH 1987).

The success of the next step, fertilization, is dependent on proper sperm capacitation. Sperm capacitation can be induced by various incubation procedures of fresh or frozen/thawed semen and is strongly species specific (BALL et al. 1983, PARRISH et al. 1986).

Capacitation renders the spermatozoa capable to undergo the acrosome reaction when approaching the oocyte. During the acrosome reaction the outer membrane of the sperm head is lost, and a specific enzyme (hyaluronidase) is released. Only acrosome reacted sperm can penetrate the zona pellucida and enter the oocyte (BRACKETT 1983, BAVISTER 1986, HUNTER 1987, FIRST & PARRISH 1987; Fig. 4). Once entered, the sperm head starts to decondense, whereas the oocyte chromosomes undergo the second meiotic division, extruding the second polar body and forming the female pronucleus. Complete male pronucleus formation is dependent on the support of factors being produced within the oocyte during cytoplasmic maturation.

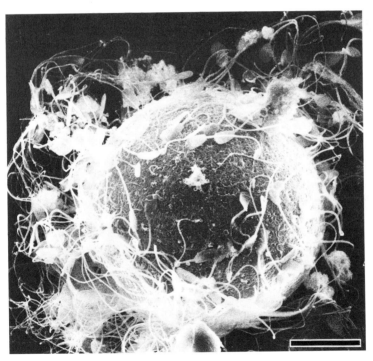

Fig. 4. In vitro matured bovine oocyte after incubation with bovine spermatozoa. Note the large number of spermatozoa sticking to the surface of the zona pellucida. (Scanning electron micrograph taken by J. Kassner, Institute of Cell Biology, ETH-Zürich). Bar = 20μm.

The two pronuclei of live bovine ova are not visible through the light microscope in contrast to pronuclei of mice or rabbit ova. By centrifugation, the obscuring granular material can be moved to the side and the pronuclei become visible (ROBL et al. 1987, WALL & HAWK 1988). No serious harm seems to result from this treatment. This trick will allow the manipulation of the bovine pronuclei. For example, pronuclear stage ova from normal animals may be used as recipients of abnormal genes, in order to preserve or test these genes.

For further development, pronuclear stage ova or early embryos have to be transferred surgically to the oviduct of synchronized recipients. To avoid surgery, *in vitro* embryonic development up to the morula or blastocyst stage would be necessary. Recently, promising results were achieved by the culture of *in vitro* fertilized ova on layers of oviductal cells. Without additional cells, the development seems to be blocked at the 8 to 16 cell stage (EYESTONE & FIRST 1986, FIRST & PARRISH 1987).

In vitro cultured morulae or blastocysts can be treated similarly as flushed embryos and either be frozen or transferred.

Conservation of genetic information (DNA) within a cell system or as a gene sequence

Aside from the conservation of a haploid or diploid genome, such as sperm or embryos, respectively, special genetic information may be preserved within somatic cell lines or as an isolated gene sequence. Two conditions have to be met for the use of these methods. First, extensive knowledge of the structure of the genome, in particular of the individual gene loci and their products, is required (STRANZINGER 1987). Second, the gene product under study will be accessible only after successful gene transfer into an embryo, and subsequent expression in the born animal (HAMMER et al. 1985).

The investigation of these two conditions has only recently been initiated. Therefore the application of DNA conservation techniques to farm animals is still very limited. It can only be considered when the level of basic research on the genome and its individual products in farm animals has reached a similar level as, for example, in human genetics. Here, various genes for particular inherited human defects have been identified. Applications have already been used in biotechnology where genetically modified microorganisms successfully produce certain hormones, and in plant production where genetic manipulation gained resistant breeds. Similar results have not yet been achived in farm animals (GELDERMANN 1987). For the preservation of genetic information in farm animals as a DNA sequence, major improvements and extensive validation of these techniques are required.

Summary

The genome and genetic anomalies of farm animals can be preserved as a genetic line in animal breeding programmes, as frozen semen, as frozen embryos, within a cell system or as a gene sequence. The genome can be manipulated in ova at the pronuclear stage. Ova can be recovered either surgically from stimulated and inseminated animals or derived from *in vitro* fertilized oocytes. Information about the genome is accessible at various stages of the preservation procedures described in this chapter.

References

ADAMS C.E. (1982): Mammalian Egg Transfer. CRC Press, Boca Raton, FL.
BALL, G.D.; LEIBFRIED, M.L.; LENZ, R.W.; AX, R.L.; BAVISTER, B.D.; FIRST, N.L. (1983): Factors affecting successful *in vitro* fertilization of bovine follicular oocytes. Biol. Reprod. **28**, 717-725.
BAVISTER, B.D. (1986): Animal *in vitro* fertilization and embryo development. In: Developmental Biology, Vol. 4. Manipulation of mammalian development (Editor: GWATKIN R.B.L.). Plenum Press, New York, NY. Pp. 81- 148.
BRACKETT, B.G. (1983): A review of bovine fertilization *in vitro*. Theriogenology **19**, 1-15.
BRACKETT, B.G.; BOUSQUET, D.; BOICE, M.L.; DONAWICK, W.J.; EVANS, J.F.; DRESSEL, M.A. (1982): Normal development following *in vitro* fertilization in the cow. Biol. Reprod. **27**, 147-158.
CHENG, W.T.K.; MOOR, R.M.; POLGE, C. (1986): *In vitro* fertilization of pig and sheep oocytes matured *in vivo* and *in vitro*. Theriogenology **25**, 146 (Abstract).

CHURCH, R.B.; SCHAUFELE, F.J.; MECKLING, K. (1985): Embryo manipulation and gene transfer in livestock. Can. J. Anim. Sci. 65, 527-537.

EPPIG, J.J.; WARD-BAIL, P.F. (1982): The mechanism of cumulus cell-oocyte uncoupling: evidenic for the participation of both, cumulus cells and oocytes. Gamete Res. 6, 145-154.

EYESTONE, W.H.; FIRST, N.L. (1986): A study of the 8- to 16-cell developmental block in bovine embryos cultured in vitro. Theriogenology 25, 152 (Abstract).

FIRST, N.L.; PARRISH, J.J. (1987): In vitro fertilization of ruminants. J. Reprod. Fert., 34 ,Suppl. 151-165.

FOOTE, R.H. (1974): Artificial Insemination. In: Reproduction in Farm Animals. (Editor: HAFEZ, E.S.E.). Lea and Febiger, Philadelphia. Pp. 409-431.

FUKUI, Y.; FUKUSHIMA, M.; ONO, H. (1983): Fertilization in vitro of bovine oocytes after various sperm procedures. Theriogenology 20, 651-660.

GARNER, D.L. (1984): In vitro methods for estimating fertilizing capacity of sperm cells. 10th Int. Congr. Anim. Reprod. A.I, 10, 9-15.

GELDERMANN, H. (1987): Methods, possible applications and consequences of gene technology in animal breeding. Züchtungskunde 59, 1-16.

HAMMER, R.E.; PURSEL, V.G.; REXROADJI, C.E.; WALL, R.J.; BOLT, D.J.; EBERT, K.M.; PALMITER, R.D.; BRINSTER, R.L. (1985): Production of transgenic rabbits, sheep and pigs by microinjection. Nature 315, 680-683.

HARTMANN, J.F. (1983): Mechanism and Control of Animal Fertilization. Academic Press, New York, NY.

HUNTER, R.H.F. (1987): The timing of capacitation in mammalian spermatozoa - a reinterpretation. Res. Reprod. 19, 3-4.

LEIBFRIED-RUTLEDGE, M.L.; CRITSER, E.S.; EYESTONE, W.H.; NORTHEY, D.L.; FIRST, N.L. (1987): Development potential of bovine oocytes matured in vitro or in vivo. Biol. Reprod. 36, 376-383.

LEIBO, S.P. (1986): Cryobiology: preservation of mammalian embryos. In: Genetic Engineering of Animals. in Agricultural Perspective. Basic Life Science, Vol. 37 (Editors: EVANS, J.W.; HOLLAENDER, A.). Plenum Press, New York, NY. Pp. 251-272.

PARRISH, J.J.; SUSKO-PARRISH, J.L.; LEIBFRIED-RUTLEDGE, M.L.; CRITSER, E.S.; EYESTONE, W.H.; FIRST, N.L. (1986): Bovine in vitro fertilization with frozen-thawed semen. Theriogenology 25, 591-600.

PAUFLER S.K. (1974): Künstliche Besamung und Eitransplantation bei Tier und Mensch. Verlag M. u. H. Schaper, Hannover, G.F.R.

RALL, W.F.; Meyer, T.K. (1989): Zone fracture damage and its avoidance during the cryopreservation of mammalian embryos. Theriogenology 31, 683-692.

ROBL, J.M.; PRATHER, R.; BARNES, F.; EYESTONE, W.; NORTHEY, D.; GILLIGAN, B.; FIRST, N.L. (1987): Nuclear transplantation in bovine embryos. J. Anim. Sci. 64, 642-647.

ROTTMANN, O. (1987): Gentechnologie in der Tierproduktion - Manipulation und Analyse des Genotyps. Habilitationsschrift der Techn. Univ. München, Landw. Fakultät Freising-Weihenstephan.

SEIDEL, G.E., Jr. (1981): Superovulation and embryo transfer in cattle. Science 211, 351-358.

STRANZINGER, G. (1987): Gene mapping and gene homologies in farm animals: techniques and present status of gene maps. Animal Genetics 18, Suppl. 1, 111-116.

SÜSS, U.; WÜTHRICH K.; STRANZINGER, G. (1988): Chromosome configurations and time sequence of the first meiotic division in bovine oocytes matured in vitro. Biol. Reprod. 38, 871-880.

WALL, R.J.; HAWK, H.W. (1988): Development of centrifuged cow zygotes cultured in rabbit oviducts. J. Reprod. Fert. 82, 673-680.

WILLADSEN, S.M. (1977). Factors affecting the survival of sheep and cattle embryos during deep-freezing and thawing. In: The Freezing of Mammalian Embryos. (Editors: ELLIOT, K.; WHELAN, J.). Elsevier Excerpta Medica North-Holland, Amsterdam/Oxford/New York. Ciba Foundation Symposium 52 (New Series), 175-194.

WISE, T.; VERNON, M.W.; MAURER R.R. (1986): Oxytocin, prostaglandins E and F, oestradiol, progesterone, sodium and potassium in preovulatory follicles either developed normally or stimulated by follicle stimulating hormone. Theriogenology 26, 757-778.

XU, K.P.; GREVE, T.; CALLESEN, H.; HYTTEL, P. (1987): Pregnancy resulting from cattle oocytes matured and fertilized in vitro. J. Reprod. Fert. 81, 501- 504.

VAN ZUTPHEN, L.F.M.; PRINS, J.B. (1988): Genetic monitoring of inbred strains: some practical considerations. 26th Scientific Meeting of the Society for Laboratory Animal Science. 125 (Abstracts).

Author's address:
U. Süss, Givaudan Forschungsgesellschaft SA, CH-8600 Dübendorf, Switzerland.

Transgenic animals as models in biomedical research

K. BÜRKI

Introduction

Animal models which closely mimic human diseases are instrumental for the understanding of disease mechanisms as well as for the development of new preventive or therapeutic treatments. Classically, such animal models have been selected from spontaneous mutants occurring in breeding colonies of laboratory animals or livestock. Today, transgenic technologies offer possibilities to change the physiology of animals and therefore to experimentally generate precise models for human genetic diseases.

Transgenic animals carry experimentally introduced cloned DNA stably integrated in their genome. The foreign DNA sequences can be functional genes, so-called transgenes, or sequences designed to interrupt the correct expression of an endogenous gene. Structural genes can be expressed in transgenic animals either under the control of their own regulatory sequences, or under the control of heterologous regulatory sequences. The regulatory sequences present, enhancers and promoters, will determine the stage and cell type specific expression of the particular structural gene. Thus, the expression of any defined gene product can be targeted to a tissue of choice (see reviews by PALMITER & BRINSTER 1986, JAENISCH 1988).

The present short review focuses on approaches to experimentally generate mouse models for human diseases. It will not attempt to give a complete list of all proposed particular models, but emphasis will rather be placed on principles and recent developments.

Methods for introducing genes into animals

Three different methods have successfully been used to transfer genes into animals: 1) the direct microinjection of recombinant DNA into one of the pronuclei of the freshly fertilized egg; 2) the transduction of foreign DNA by retroviruses or retroviral vectors into embryos at various stages of development; and 3) the use of genetically transformed embryonic stem cells as vehicles.

The direct microinjection of a few hundred copies of the recombinant transgene into one of the pronuclei of the fertilized egg is currently the most widely applied method for introducing genes into animals (Fig. 1). The main advantage of this method is its reproducible efficiency and its applicability to many mammalian species. With mice, the overall efficiency (transgenic animals/eggs injected) is in the range of a few percent (BRINSTER et al. 1985). The same holds true for rats (unpublished results). After microinjecting eggs of larger farm animals the overall efficiency usually is somewhat lower (about one percent), probably mainly due to difficulties in visualizing the pronuclei and/or due to suboptimal in vitro culture conditions. Nevertheless, several investigators have successfully produced transgenic rabbits (BREM et al. 1985, HAMMER et al. 1985, BÜHLER et al. 1987), sheep (HAMMER et al. 1985, SIMONS et al. 1988) and pigs (BREM et al. 1985, HAMMER et al. 1985).

After microinjection of cloned DNA into one of the pronuclei usually multiple DNA copies integrate in a head-to-tail arrangement at one random chromosomal site into the host genome. If integration takes place into the genome of the zygote, the resulting animal will carry copies of the transgenes in all its cells, including the germ cells. The stably integrated copies of newly acquired DNA are present in a hemizygous condition and are therefore transmitted to fifty percent of the offspring of a transgenic founder animal. By interbreeding hemizygous first generation carrier siblings, homozygous transgenic animals may be obtained in the second generation. Occasionally, homozygous animals cannot be obtained, because the integration of the transgenes has interrupted an endogenous gene essential for the normal embryonic or fetal development (GRIDLEY et al. 1987). If the integrated transgenes comprise functional regulatory sequences, they will be expressed. Phenotypical effects of the transgene expression are inherited as dominant traits.

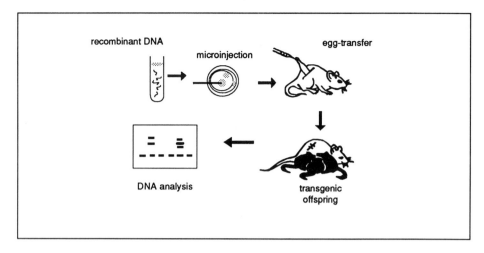

Fig. 1. Schematic outline of the production of transgenic mice by pronuclear microinjection.

Although technically simple, the use of retroviruses and retroviral vectors has not found widespread application. This is mainly due to the size limitations for transduced DNA as well as to the unresolved problems of reproducibly expressing a transduced eukaryotic gene.

A rapidly evolving method to produce transgenic mice makes use of embryonic stem cells as carriers of recombinant DNA into the early embryo (Fig. 2). Embryonic stem cell lines are derived from explanted blastocysts. They retain their normal karyotype and their pluripotent embryonic character even after DNA transfection (viral transduction, calcium phosphate precipitation, electroporation or microinjection). When such cells are placed back into a carrier blastocyst, they can colonize the developing embryo and contribute to the germ-line of a resulting chimeric organism (GOSSLER et al. 1986). Functional embryonic stem cell-derived germ cells will transmit the *in vitro* mutated genome to the next generation. So far, embryonic stem cells have been derived from mice and hamsters only. Therefore, this approach is not yet applicable to other mammalian species.

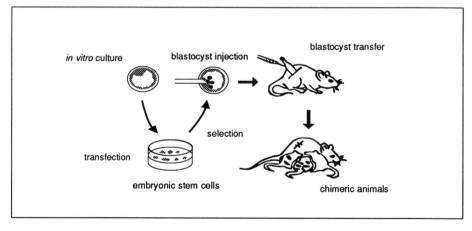

Fig. 2. Schematic outline of the introduction of selected mutations into an organism by the use of blastocyst derived embryonic stem cells.

Disease models

Various approaches have been used to experimentally generate potential disease models. These are, together with prominent examples, listed in Table 1.

The most straightforward approach is the expression of a transgene in order to cause an elevated level of the gene product either in the circulating plasma or in a particular cell type. Its most widely used application consists in the tissue specific expression of oncogene encoded regulatory proteins, which has led to a variety of transgenic mouse lines exhibiting the constitutive development of particular tumors (reviewed by PALMITER & BRINSTER 1986, JAENISCH 1988). One such experimental mouse model (LEDER et al. 1986) has recently been patented in the United States.

Table 1. Mouse models for human diseases generated by transgenic technologies.

approach	gene(s)	model	references
tissue specific expression	oncogenes	tumor development	JAENISCH (1988) PALMITER & BRINSTER (1986) (reviews)
overexpression	Cu/Zn superoxide dismutase gene	Down syndrome	EPSTEIN et al. (1987)
expression of receptor in liver	low density lipoprotein (LDL) receptor gene	low plasma LDL levels	HOFMANN et al. (1988)
expression of mutated gene	mutated alpha1(I) collagen gene	osteogenesis imperfecta	STACEY et al. (1988)
expression of antisense RNA	myelin basic protein antisense minigene	reduced myelination	KATSUKI et al. (1988)
cell ablation: tissue specific expression	diphteria toxin gene (subunit A)	microphthalamia pancreatic rudiment dwarfism	BREITMANN et al. (1987) PALMITER et al. (1987) BEHRINGER et al. (1988)
selection of deficiency mutants of embryonic stem cells	hypoxanthinphosphoribosyltransferase (HPRT) gene	(Lesch-Nyhan)	HOOPER et al. (1987) KUEHN et al. (1987)

Similarly, by introducing the human Cu/Zn-superoxide dismutase gene into the genome of mice, the level of functional enzyme could be increased up to six fold in the brain and other tissues (EPSTEIN et al. 1987). Such transgenic mice may therefore provide insight into the consequences of increased dosage of the Cu/Zn-superoxide dismutase gene in Down syndrome.

An interesting approach has been chosen by HOFMANN et al. (1988). In order to reduce plasma levels of low density lipoproteins (LDL), these authors expressed the LDL-receptor on the surface of liver parenchymal cells by using the metallothionein-promoter. Induction of this gene construct by heavy metals (intraperitoneal injection or in drinking water) caused a high density of LDL receptors on liver cells, subsequently followed by a drop of the level of free plasma LDL. These observations suggest that overexpression of other receptors, for example insulin- or transferrin-receptors, might have similar effects and might deprive peripheral organs of the ligand.

The severe, inherited human disease perinatal lethal osteogenesis imperfecta has been reproduced in mice by inserting and expressing a mutated pro-alpha1(I) collagen gene (STACEY et al. 1988). The expression of as little as 10% mutant collagen in transgenic fetuses resulted in a dominant lethal phenotype due to a reduced type I collagen content. This observation suggests that the presence of nonfunctional subunits may severely inhibit the correct assembly of multimeric structures. This should allow the study of the consequences of inappropriate assembly of other collagens or of cytoskeleton proteins, such as actins or intermediate filaments.

Although antisense RNA (minus strand RNA complementary to messenger RNA) has been shown to effectively repress the expression of specific genes of *in vitro* cultured cells at the translational level, with transgenic mice only one successful attempt has been reported: by coexpressing an antisense basic myelin protein mini-gene in myelin producing cells, the myelination in the central nervous system of the transgenic mice was significantly reduced, resulting in shiverer phenotypes (KATSUKI et al. 1988). Using strong promoters, this approach may be used to reduce the translational efficiency of any characterized messenger RNA.

A yet different approach makes use of the cell toxicity of the diphteria toxin subunit A. By fusing the structural gene for this toxin subunit to a cell-type specific promoter, it should be possible to induce autonomous death of all cells in which that specific promoter is active. This targeted cell ablation, also named toxigenics (BEDDINGTON 1988), has been applied to generate transgenic mice exhibiting microphthalmia (reduced lens mass), pancreatic rudiments or dwarfism, depending on the tissue specificity of the promoters used: gamma-crystallin promoter, elastase I promoter, and growth hormone promoter, respectively (for references see Table 1). An incomplete penetrance of toxicity and a highly variable expression of the toxin gene constructs make interpretations difficult. However, certain pathological conditions where particular cell types are reduced or absent may well be mimicked by genetic cell ablation in transgenic animals.

Using mutant embryonic stem (ES) cells, two groups have reported the successful introduction into mice of deficiencies for the X-linked gene encoding hypoxanthin-phosphoribosyltransferase (HPRT) (HOOPER et al. 1987, KUEHN et al. 1987). A deficiency of this enzyme in humans causes severe neurological disorders, described as Lesh-Nyhan syndrome. Although the syndrome is well understood in terms of molecular genetics, little is known about how the disorder develops; no preventive treatment is available. Surprisingly, the HPRT-deficient mice do not represent a model for the human disease, because mice seem to use different pathways of purine metabolism and do not exhibit the typical neuronal damage observed in human patients. Nevertheless, these experiments clearly demonstrate the potential of the use of ES cells for the introduction of specific mutations into mice (for reviews see HOGAN 1987, JACKSON 1987).

Conclusions

The generation of transgenic animals by microinjection of recombinant DNA sequences into one of the pronuclei of the fertilized egg will continue to be the method of choice, if simple expression of a transgene added to the genetic background of the host mammal is expected to affect its physiology in the desired way. This approach is straightforward and is applicable to a variety of species. Moreover, by expressing antisense RNA the expression of a particular gene may be down modulated. Specific expression of a toxin gene may ablate selected cells. Expression of a non-functional subunit may drastically reduce the level of intact multimeric structures. Therefore, a variety of different approaches may be considered for the generation of a potential animal model for a human disease.

More refined models, however, will be generated by using genetically transformed ES cells. An advantage of this methodology is the possible selection of appropriately transformed cells prior to the introduction into a host embryo. In addition, the use of ES cells is attracting increasing attention because they allow the application of methods to target mutations to selected genes by homologous recombination. Recently developed ingenious selection systems (MANSOUR et al. 1988) and powerful screening methods (SAIKI et al. 1988) make it possible to rescue and identify the few rare cells which have undergone homologous recombination of an endogenous target gene with a transfected mutant copy. With these methods it is therefore possible to precisely modify a given gene or to insert sequences which interrupt its expression. As mentioned above, selected mutant cells can be introduced into a developing host embryo and populate the germline of a resulting chimeric animal. In this way an experimentally introduced mutation can be transmitted to the offspring of such a chimera and subsequently be stabilized in a mutant mouse line. By crossing to homozygosity the effect of an induced recessive mutation, for example a deficiency of a selected autosomal gene, will be revealed. This multi-step approach is currently restricted to mice (and hamsters), but efforts to establish *in vitro* embryonic stem cells of various other mammalian species may soon be successful.

The near future will see a wealth of new experimentally generated mouse mutants. Their impact on our understanding of normal development as well as of disease mechanisms will be dramatic.

References

BEDDINGTON, R.S.P. (1988): Toxigenics: strategic cell death in the embryo. Trends Genet. **4**, 1-2.

BEHRINGER, R.R.; MATHEWS, L.S.; PALMITER, R.D.; BRINSTER, R.L. (1988): Dwarf mice produced by genetic ablation of growth hormone-expressing cells. Gene. Develop. **2**, 453-461.

BREITMAN, M.L.; CLAPOFF, S.; ROSSANT, J.; TSUI, L.-C.; GLODE, L.M.; MAXWELL, I.H.; BERNSTEIN, A. (1987): Genetic ablation: targeted expression of a toxin gene causes microphthalmia in transgenic mice. Science **238**, 1563-1565.

BREM, G.; BRENIG, B.; GOODMAN, H.M.; SELDEN, R.C.; GRAF, F.; KRUFF, B.; SPRINGMAN, K.; HONDALE, J.; MEYER, J.; WINNACKER, E.-L.; KRÄUSSLICH, H. (1985): Production of transgenic mice, rabbits and pigs by microinjection into pronuclei. Zuchthygiene **20**, 251-252.

BRINSTER, R.L.; CHEN, H.Y.; TRUMBAUER, M.E.; YAGLE, M.K.; PALMITER, R.D. (1985): Factors affecting the efficiency of introducing foreign DNA into mice by microinjecting eggs. Proc.Natl.Acad.Sci.USA **82**, 4438-4442.

BÜHLER, T.A.; REDING, T.; WENT, D.F.; WEILENMANN, R.; FRIES, R.; STRANZINGER, G. (1987): Microinjection of mouse IgM-genes into pronuclei of rabbit eggs. Theriogenology **27**, 216.

EPSTEIN, C.J.; AVRAHAM, K.B.; LOVETT, M.; SMITH, S.; ELROY-STEIN, O.; ROTMAN, G.; BRY, C.; GRONER, Y. (1987): Transgenic mice with increased Cu/Zn-superoxide dismutase activity: animal model of dosage effects in Down syndrome. Proc.Natl.Acad.Sci. USA **84**, 8044-8048.

GOSSLER, A.; DOETSCHMAN, T.; KORN, R.; SERFLING, E.; KEMLER, R. (1986): Transgenesis by means of blastocyst-derived embryonic stem cell lines. Proc.Natl.Acad.Sci.USA **83**, 9065-9069.

GRIDLEY, T.; SORIANO P., JAENISCH, R. (1987): Insertional mutagenesis in mice. Trends Genet. **3**, 162-166.

HAMMER, R.E.; PURSEL, V.G, REXROAD, C.E.; WALL, R.J.; BOLT, D.J.; EBERT, K.M.; PALMITER, R.D.; BRINSTER, R.L. (1985): Production of transgenic rabbits, sheep and pigs by microinjection. Nature **315**, 680-683.

HOFMANN, S.L.; RUSSELL, D.W.; BROWN, M.S.; GOLDSTEIN, J.L.; HAMMER, R.E. (1988): Overexpression of low density lipoprotein (LDL) receptor eliminates LDL from plasma in transgenic mice. Science **239**, 1277-1281.

HOGAN, B. (1987): Lesch-Nyhan syndrome: engineering mutant mice. Nature **326**, 240-241.

HOOPER, M.; HARDY, K.; HANDYSIDE, A.; HUNTER S., MONK, M. (1987): HPRT-deficient (Lesch-Nyhan) mouse embryos derived from germline colonization by cultured cells. Nature **326**, 292-295.

JACKSON, I.J. (1987): The real reverse genetics: targeted mutagenesis in the mouse. Trends Genet. **3**, 119-120.

JAENISCH, R. (1988): Transgenic animals. Science **240**, 1468-1474.

KATSUKI, M.; SATO, M.; KIMURA, M.; YOKOYAMA, M.; KOBAYASHI, K.; NOMURA, T. (1988): Conversion of normal behavior to shiverer by myelin basic protein antisense cDNA in transgenic mice. Science **241**, 593-595.

KUEHN, M.R.; BRADLEY, A.; ROBERTSON, E.J.; EVANS, M.J. (1987): A potential animal model for Lesch-Nyhan syndrome through introduction of HPRT mutations into mice. Nature **326**, 295-298.

LEDER, A.; PATTENGALE, P.K.; KUO, A.; STEWART, T.A.; LEDER, P. (1986): Consequences of widespread deregulation of the c-myc gene in transgenic mice: multiple neoplasms and normal development. Cell **45**, 485-495.

MANSOUR, S.L.; THOMAS, K.R.; CAPECCHI, M.R. (1988): Disruption of the proto-oncogene int-2 in mouse embryo-derived stem cells: a general strategy for targeting mutations to nonselectable genes. Nature **336**, 348-352.

PALMITER, R.D.; BEHRINGER, R.R.; QUAIFE, C.J.; MAXWELL, F.; MAXWELL, I.H.; BRINSTER, R.L. (1987) : Cell lineage ablation in transgenic mice by cell-specific expression of a toxin gene. Cell **50**, 435-443.

PALMITER, R.D.; BRINSTER, R.L. (1986) : Germ-line transformation of mice. Ann. Rev. Genet. **20**, 465-499.
SAIKI, R.K.; GELFAND, D.H.; STOFFEL, S.; SCHARF, S.J.; HIGUCHI, R.; HORN, G.T.; MULLIS, K.B.; ERLICH, H.A. (1988): Primer directed enzymatic amplification of DNA with a thermostable DNA polymerase. Science **239**, 487-491.
SIMONS, J.P.; WILMUT, I.; CLARK, A.J.; ARCHIBALD, A.L.; BISHOP, J.O.; LATHE, R. (1988): Gene transfer into sheep. Bio-Technology **6**, 179-183.
STACEY, A.; BATEMAN, J.; CHOI, T.; MASCARA, T.; COLE, W.; JAENISCH R. (1988): Perinatal lethal osteogenesis imperfecta in transgenic mice bearing an engineered mutant pro-alpha1(I) collagen gene. Nature **332**, 131-136.

Author's address:
K. Bürki, Preclinical Research, Sandoz Ltd., CH-4002 Basel, Switzerland.

Index

A

abortion 17, 43
acetylcholine 103
acid phosphatase 26
acrosome reaction 173
ACTH
    see adrenocorticotropic hormone
adaptation 57
adaptation phase
    see general adaptation syndrome
adenohypophysis 62, 103
adenosine monophosphate, cyclic (cAMP) 11, 61, 64
    cAMP-phosphodiesterase 11
adipose tissue 63
adrenal
    cortex 61, 81, 104
    gland 69
    medulla 60, 61
    zona fasciculata 62
    zona glomerulosa 62
    zona reticularis 62
adrenaline 97
adrenocorticotropic hormone (ACTH) 61, 62, 63, 64, 69, 73, 81, 98, 103, 104
    circadian rhythmicity 65
    release 72, 73
    variants 82, 83, 84, 85
aestivation 63
aggression 97
α-agonist (adrenergic)
    noradrenaline 61
β-agonist (adrenergic)
    isoproterenol 61
alarm reaction
    see general adaptation syndrome
aldosterone 62
alloimmunization 137
allotype 137
amino acid
    excitatory contents 97
    substitution 158
    transport 64
γ-aminobutyric acid 104
aminotransferase 63
amphetamine 100
amygdala 102
analgesia 103
anaphase 50
anesthesia 65
aneuploidy 46, 47, 48, 49, 53
    mosaicism 46
angiotensin 62
animal
    transgenic 17
animal model 9

animal model (cont.)
    cancer 9
    chicken 43
    goat 147
    inflammation 9
    mouse 116, 177
    pig 69, 81, 89, 111, 131
    rabbit 33
    sheep 155, 161
    stress 65
anterior pituitary
    see adenohypophysis
antibody
    antiviral 150
antihemophilic drugs 161
antihemophilic factor
    see factor VIII
antiinflammatory effect 63
anxiety 97, 101
Apo-B
    see apolipoprotein B
apolipoprotein B (Apo-B) 134
arthritis 151
    mycoplasma induced 151
    resistance/susceptibility 151
    rheumatoid-like 151
    type II, collagen induced 151
artificial
    insemination 167
    vagina 168
ataxia 148
atherosclerosis 131
    dietary induced 133
    pig 132
autosome 48, 53

B

banding 19, 23
    C-, G-, Q-, R-banding 44
behavior
    avoidance 98, 100, 104
    freezing 98, 103
    hyperactivity 104
    model 102
    motor 99
    reproductive 21
    stressor induced 102
bile acid excretion 36
biotin labeling 23
blastocyst 171
blastoderm 45, 49
bleeding 155, 161
blood group 21
brachygnathia 12
Brattleboro rat 9, 10
BrdU
    see 5-bromodeoxy-uridine
breeding programme 167
5-bromodeoxy-uridine 19, 43

bursa Fabricii 45

C

C-band 19, 53
CAE
    see caprine arthritis encephalitis
CAE virus
    see virus, caprine arthritis encephalitis
cAMP
    see adenosine monophosphate, cyclic
cAMP-phosphodiesterase 12
caprine arthritis encephalitis (CAE) 147
caprine leukocyte antigen (CLA) 150
capture myopathy 63
carbohydrates
    synthesis 63
cardiovascular functions 102
carpitis 147
CAT box 18
catecholamines 60, 64, 89, 92
    adrenaline 73, 91
    dopamine 91
    intoxication 63
    metabolites 63
    noradrenaline 73, 91
cattle
    idiogram 25
    insemination 168
caudate nucleus 99
CBG
    see cortisol binding globulin
cell
    ablation 180
    hybridization 23
centric fusion 53
centromere 49, 53
    regions 19
cervix uteri
    species differences 168
chicken
    cytogenetic studies 44
    embryo, heteroploid 45
chimera 53, 178
chimeric organism 179
chimerism
    diploid/diploid (2N/2N) 48
    diploid/triploid (2N)/(3N) 47
    haploid (1N) 49
    haploid/diploid (1N/2N) 49
cholesterol
    absorption 34
    de novo synthesis 34, 35
    diets 39, 40, 134
    homeostasis 34
    hyperresponder 38
    hyporesponder 38
    in blood 33, 37
    inter-species differences 37
    intra-species differences 37

cholesterol *(cont.)*
    metabolism 33
cholineacetyl transferase 102
chromosome
    aberration 43, 50
    abnormal complement 43
    abnormality 44, 48
    analysis 44
    banding 23, 45
    banding homology 26
    inversion 50
    lagging 53
    marker 20, 44
    marker production 46
    microchromosome 44
    morphology 20
    multivalent 54
    numbers 20
    pericentric inversion 46
    rearrangement 50
    reciprocal translocation 46
    segregation 50
    submetacentric 26
    telocentric 19
    translocation 50
chylomicron 34
CLA
    see caprine leukocyte antigen
CLIP
    see corticotropin-like intermediate lobe peptide
coagulation
    cascade 156, 157
    factor 155
    factor inhibitors 160
    factor, vitamin K dependent 157
coagulopathies 155
cognitive process 101
colchicine 45
collagen 155
    gene 179, 180
colostrum 149
    CAE free 150
coronary heart disease 33
corpus luteum 169
cortex
    prefrontal 99, 101, 102
corticosteroids 70, 97, 98, 103
    cortisol 63, 69, 72, 81
    corticosterone 61, 98, 103
corticotropic activity 84
corticotropic cells 103
corticotropin releasing factor (CRF) *or* corticotropin
    releasing hormone (CRH) 62, 63, 69, 103, 104
corticotropin-like intermediate lobe peptide (CLIP)
    81
cortisol 63, 69, 81
    release 72
cortisol binding globulin (CBG) 70

Coturnix
    see quail
CRF
    see corticotropin releasing factor
CRH
    see corticotropin releasing factor
cumulus cells 172
cytokinesis 49
cytometry 45

D

DA
    see dopamine
DDAVP
    see vasopressin, deamino-D-arginine vasopressin
deoxycorticosterone acetate (DOCA) 9, 11
diabetes insipidus 10
    nephrogenic (NDI) 11
diabetes mellitus 9
diagnosis
    prenatal 17
diakinesis 44
diencephalon 60
dietary
    fatty acids 39, 134
    proteins 40
digyny 53
    oocyte 46
diphtheria toxin 179, 180
disease model
    artificial 10
disease resistance 66
disjunction 50
dispermy 47, 53
distance
    genetic 20
    morphologic 20
DNA
    conservation 174
    recombinant 177
    transfection 177
    viral 150
DOCA
    see deoxycorticosterone acetate
donor 171
DOPAC
    see dopamine metabolites
dopamine (DA) 97, 99, 102
    metabolism 101
    turnover 99
dopamine metabolites 101
    dihydroxyphenylacetic acid (DOPAC) 99, 102, 103
    homovanillic acid (HVA) 99, 102, 103
Down's syndrome 179
dummy mounting 94
dwarfism 179, 180

E

egg 43, 44
    yolk 45
ejaculate 168
ELISA 149, 150
embryo 43, 44
    androgenetic haploid 47
    chimeric 47
    diploid/triploid (2N/3N) 47
    euploid chimeric 47
    freezing 170
    haploid/diploid (1N/2N) 47
    haploid/triploid (1N/3N) 47
    mortality 43, 51
    mosaic 47
    pure haploid 47
    splitting 170
    stem cells 177
    storage 170
    thawing 170
    transfer 168, 170, 178
emotional factors 98
emotionality 102
EMS
    see radiomimetic agents
endonuclease 18
    restriction enzyme 18
endoproteolysis 81
entropium congenitum 12
    sheep 13
environment
    gene frequencies 17
    genetic interactions 150
    novel situation 101, 103
    social, stress 58, 59
enzyme defects 11
epinephrine
    see adrenaline
epitope 137
ethology 66
euploidy 46, 53
    chimeras 44, 46
exhaustion phase
    see general adaptation syndrome
exon 18
exploratory drive 98
eye field
    frontal 101

F

F VIII
    see factor VIII
F IX
    see factor IX
factor VIII (F VIII) 11, 155, 156, 157, 161
    concentrates 158

factor VIII (F VIII) deficiency
     see hemophilia
factor IX (F IX) 11
fatty acids
     dietary 39, 134
fear 103
feedback 97, 104
fertilization
     in vitro 171
Feulgen reagent 19
fibrin 155
fibroblasts 148
flight-and-fight reaction 63
follicle stimulating hormone (FSH) 26, 169
food deprivation 63
footshock 101, 104
FSH
     see follicle stimulating hormone

G

G-6-P
     see glucose-6-phosphate dehydrogenase
G-band 19, 26, 54
G-blood group 21
gametes 43, 44
GAS
     see general adaptation syndrome
gastrointestinal tract
     disorders 57
gene
     aberration 17
     acid phosphatase 2 (ACP2) 26
     basic myelin protein 179, 180
     collagen 179, 180
     factor VIII 158
     follicle stimulating hormone (FSH) 26, 27
     halothane 12
     hemoglobin beta (HBB) 26, 27
     homology 23, 26
     hypoxanthin phosphoribosyl transferase (HPRT) 179
     keratin 26, 27
     lactate dehydrogenase A (LDHA) 26
     localization on chromosomes 20
     mapping 17, 18, 167
     oncogene 179
     parathyroid hormone (PTH) 26, 27
     product 17
     stress linked 66
     superoxide dismutase 179
     swine lymphocyte antigen (SLH) 27
     transfer 20
general adaptation syndrome (GAS) 57, 58
genetic
     anomalies 167, 170
     diseases 17
     distance 20
     diversity 18
     engineering 18

genetic (cont.)
     information 18, 19
     marker 44, 167
genome
     evolutionary changes 17
     map 17
gestation 61
GH
     see growth hormone
glucagon 64
glucocorticoids 61, 73, 90, 102
     11-desoxycortisol 62
     corticosterone 62
     cortisol 62
     cortisone 62
gluconeogenesis 64
glucose
     metabolism 63
     tolerance 71, 72
     transport 64
     utilisation 63
glucose-6-phosphate dehydrogenase (G-6-P) 64
glycogen synthetase 64
glycogen-glucose equilibrium 62
glycogenolysis 61, 64
glycolytic enzymes 63
GnRH
     see gonadotropin releasing hormone
goat 147, 168
     caprine arthritis encephalitis (CAE) 147
     Saanen 151
     Toggenburg 150
gonadotropin releasing hormone (GnRH) 65
gonocyte 43
gonosomes 43, 54
growth hormone (GH) 73

H

haemophilia
     see hemophilia
halothane
     gene 12
     reaction 69
haploid 54
haplotype 151
HBB
     see gene, hemoglobin beta
HDL
     see lipoprotein, high density
hemophilia
     genetics 158, 160
     history 156, 157
     in animals 159
     in humans 155
     pathology 155
     therapy 158, 160
hemostasis 155
heterochromatin 43, 45, 54
heteroploidy 45, 46, 47, 48, 54

heterozygosity 50
hibernation 63
high performance liquid chromatography (HPLC) 82
hippocampus 102
histone proteins 19
homeostasis
    mechanism 57
HPLC
    see high performance liquid chromatography
HPRT
    see gene, hypoxanthin phosphoribosyl transferase
human genome 17
HVA
    see dopamine metabolites
hybridization 23
    in situ 23, 26, 27, 150
5-HIAA
    see serotonin, metabolite
17-α-hydroxylase 62
5-hydroxytryptamine (5-HT)
    see serotonin, metabolite
5-HT
    see serotonin
Hyp/Y mouse
    see rickets, X-linked hypophosphatemic
hypercholesterolemia
    atherosclerosis 134
    coronary artery 140
hyperlipidemia 134
hyperlipoproteinemia 136, 139
    Zucker rat 9
hypertension
    malignant 11
    rat 9
hyperthermia
    malignant 11, 69
hypodiploidy 54
hypophysiotropic effect 62
hypothalamo-pituitary-adrenocortical axis 61, 103
    feedback control 63
hypothalamus 99, 102, 103

I

idiogram
    cattle 25
    sheep 25
IgM
    see rheumatoid factor
immobilization 101
immunogenetic 131, 137, 138
immunosuppression
    drug-induced 149
immunosuppressive effect 63, 66
in situ
    see hybridization
in vitro
    fertilization 171

in vitro (cont.)
    maturation 173
inappetentia 57
inbred line 44
inflammation
    mononuclear 148
inflammatory response 63
infrapyramidal fiber projection 104
inheritance
    Mendel's rules 17
insulin 73, 85
    resistance 71, 74, 75
interferon 149
interleukin-1 149
interstitial pneumonia 148
intron 18
inversion 54

K

karyotype 19, 20
    cattle 22, 24
    chicken 43
    fowl 22
    goat 22
    horse 22
    human 21
    pig 22
    sheep 22, 24
keratin B gene 26
Krebs cycle 64

L

lactate dehydrogenase 26
LDHA
    see gene, lactate dehydrogenase A
LDL
    see lipoprotein, low density
lecithin:cholesterol transferase 142
Lesh-Nyhan syndrome 179, 180
LH
    see luteinizing hormone
lipolysis 62, 63
lipolytic activity 84
lipoprotein
    high density (HDL) 33
    lipoprotein(a) 142
    low density (LDL) 33, 134, 179
    very low density (VLDL) 33
loading 92, 93
locomotion 99
lod score test 21
luteinizing hormone (LH) 169
lymphocyte 45, 63
lymphoma
    malignant 151

M

macrochromosome 44, 48
macrophage 148
major histocompatibility complex (MHC) 26, 150, 152
malignant
    hypertension 11
    hyperthermia 11, 69
    lymphoma 151
    tumor 63
marker
    assisted selection 17
    chromosome 44, 46, 47, 54
    genetic 44
mastitis 147
mating 92, 93
maturation
    cytoplasmic 173
    in vitro 171
    meiotic 172
meat
    pale soft exudative (PSE) 69
medulla oblongata 60
meiosis 18, 43
α-melanocyte stimulating hormone (αMSH) 81
Mendel
    rules of inheritance 17
mesocortical projection
    see cortex, prefrontal
mesohippocampus 102
mesolimbic system 99
metabolic defects 11
metacentric element 43
metaphase 19, 23
    meiotic 43
    mitotic 43
7-methylguanosine
    see mRNA
MHC
    see major histocompatibility complex
micrognathia 12
    sheep 14
microinjection 177, 178
microphthalmia 179, 180
midbrain 103
milk
    CAE free 150
    components 17
    substitutes 40
mineralocorticoids 62
mitosis 43, 49
monocyte 148
monosomy 54
morula 171
mosaic 54
mosaicism
    2N/4N 49
    diploid/hypodiploid 49
mouse
    nude 9
    transgenic 178
mRNA
    see RNA, messenger RNA
    7-methylguanosine 18
αMSH
    see melanocyte stimulating hormone
multiple sclerosis 148
mutagenesis 23
mutagens 19
mutation
    autosomal recessive 17
    de novo 19
    deletion 158
    mapping 21
    point 158
    recessive 17
    sex linked 17
    single amino acid substitution 85
    somatic 17
    X-chromosome, deletion 22, 23
myopathy, capture 63

N

NDI
    see diabetes insipidus
neocortex 60
neoplastic processes 66
nervous system
    autonomic 63
neuroendocrine functions 102
neurohypophyseal hormones 60
neurohypophysis 61
neurons
    cholinergic 99, 102
    dopaminergic 104
    monoaminergic 97
    neurosecretory 62
    peptidergic 99
    serotonergic 97, 99, 102
neuropeptide 85
nigrostriatal system 100
nigrostriatum 99
Nobel-Collip drum 9
noise 101
non-histone proteins 19
nondisjunction 54
    sex chromosomes 48
NOR
    see nuclear organizing region
noradrenaline 97
    see also α-agonist
norepinephrine
    see noradrenaline
noxious stimuli 57
nuclear organizing region (NOR) 54
nucleus, caudate 102

O

obesity 74
oestrus 169
olfactory tubercles 99
oocyte 169, 172
    binucleated 48
osteogenesis imperfecta 179, 180
oviduct 169, 174
ovulation 169
oxytocin 61

P

pain 65, 103
palate deformation
    brachygnathia 12
    micrognathia 12, 14
    Pierre-Robin syndrome 12
parvicellular nuclei 62
PCR
    see polymerase chain reaction
pentaploidy (5N) 47, 54
phenylketonuria 17
phyletic memory 59
Pierre-Robin syndrome 12
pig
    atherosclerosis 131
    G-blood group 21
    halothane reaction 69
    hyperthermia 69
    insemination 168
    lipoprotein polymorphism 131, 137
    non-obese 71, 72
    noradrenaline release 92
    obese 71, 72
    porcine stress syndrome 69, 86
    stress 69, 89
    stress hormone variants 81
pituitary 69
    anterior 81
    extracts 82
    gland 81
plasminogen 142
PMSG
    see pregnant mare serum gonadotropin
pneumonia, interstitial 148
polyadenine tail 18
polymerase chain reaction (PCR) 150
polymorphism
    chromosomes 19
    low density lipoprotein (LDL) 137
    restriction fragment length polymorphism (RFLP) 139
polyploidy 46, 54
POMC
    see proopiomelanocortin
portal circulation, hypothalamic 62
post-translational processing 81

posterior pituitary
    see neurohypophysis
precursor protein 81
pregnancy 43
    rate 168
pregnant mare serum gonadotropin (PMSG) 169
prenatal diagnosis 17
preservation
    animal model 167
    rare animals 167
primates 101
proinsulin 85
    proteolytic cleavage 85
prolactin 98
promotor 18
    metallothionein 179
pronucleus 47, 171, 177
    formation 173
    manipulations 173
    species differences 173
proopiomelanocortin (POMC) 81
prostaglandin $F_{2\alpha}$ 169
protein
    catabolism 63
    dietary 40
    histone 19
    non-histone 19
proteolysis 63
PSE
    see meat, pale soft exudative
pseudo vitamin D-deficiency (PVDR)
    human 118, 121
    pig 121, 123
    symptoms 121
PTH
    see gene, parathyroid hormone
PVDR
    see pseudo vitamin D-deficiency
    see also rickets, pseudo vitamin D-deficiency

Q

Q-band 19, 54
quadriplegia 148
quail 44, 48

R

R-band 43, 54
radiomimetic agents
    ethyl methane sulfonate (EMS) 46
    triethylene melamine (TEM) 46
rat
    Brattleboro 9, 10
    hypertension 10
    nephrogenic diabetes insipidus (NDI) 11
    Roman high-avoidance (RHA/Verh) 98, 102, 103
    Roman low-avoidance (RLA/Verh) 98, 102, 103
    spontaneous hypertensive (SH) 9

rat (cont.)
    stress-susceptible 104
    Zucker 9
receptor
    benzodiazepine 99
    corticosterone 104
    imipramine 99
    low density lipoprotein (LDL) 179
    opioid 99
recipient 169, 170, 171
recombinant
    technique 17
recombinant DNA 177
renin-angiotensin system 62
replacement therapy 17, 158
resistance phase
    see general adaptation syndrome
restriction enzyme 18
retrovirus 152
RFLP
    see polymorphism
RHA/Verh
    see rat
rheumatoid arthritis 147
rheumatoid factor (IgM) 152
rickets
    pseudo vitamin D-deficiency (PVDR) 111
    symptoms 111, 117
    treatment 124
    vitamin D-resistant 111
    X-linked hypophosphatemic, human (XLH) 111, 112
    X-linked hypophosphatemic, mouse (Hyp/Y) 116
RLA/Verh
    see rat
RNA
    antisense 179, 180
    messenger RNA (mRNA) 18, 81
    viral 150
rooting 98

S

semen
    see also sperm
    fresh 168
    frozen/thawed 168
septo-hippocampal system 103
septum
    lateral 99, 104
serotonin (5-HT) 100, 103
    metabolite, 5-hydroxyindolacetic acid (5-HIAA) 100
sex chromosomes 43, 47
    aneuploidy 44
    nondisjunction 48
sexual hormones
    androgens 62
sexual interactions 66

SH
    see rat, spontaneous hypertensive
sheep
    hemophilia A 159
    idiogram 25
shock 100
    freezing behavior 98
    repeated 101
shock-stressor 100
shuttlebox 100, 104
sickle cell anemia 11
silver staining 19
single sperm fertilization 47
SLH
    see swine lymphocyte antigen
somatomotor reflex 60
SP
    see substance P
sperm
    see also semen
    capacitation 173
    irradiation 21, 46
spermatocyte 45
spermatogon 45
splitting 170
stain
    5-bromodeoxy-uridine (BrdU) 19, 43
    Feulgen reagent 19
    Giemsa 19, 54
    orcein 19
    quinacrine 54
stem cells
    embryonic 177
steroidogenic potency 84
stimuli
    noxious 57
stress 57
    see also general adaptation syndrome
    behavioral aspects 97
    cardiovascular changes 60
    genetic background 97
    genetically determined 70
    hormonal aspects 97
    neurochemical aspects 97
    noradrenaline release 92, 93
    pig 69
    porcine stress syndrome 86
    quantification 89
    restraint 65
    syndrome 89
    transport 94
stressor 69, 97
    chemical 57
    environmental 104
    injury 57
    isolation 58, 59
    overpopulation 58, 59
    physical 57
    psychosocial 66

stressor *(cont.)*
  response 58
  social environment 58, 59
striatum 101
  dorsal 99
  ventral 99
submetacentric element 43
substance P (SP) 100
substantia nigra 99
superovulation 169
swine
  see pig
swine lymphocyte antigen (SLA) 26
synaptonemal complex 50, 54
synchronization, reproductive cycle 170
syndrome
  Down's 179
  general adaptation syndrome (GAS) 57
  Lesh-Nyhan 179, 180
  Pierre-Robin 12
  porcine stress 86, 89
synovitis 152
synteny 20

T

tail pinch 101
TATA box 18
TEM
  see radiomimetic agents
testis 45
tetraploidy (4N) 54
thalamus 60
thymus
  atrophy 63
TNF
  see tumor necrosis factor
transgene 23, 177
  disease model 179
transgenic animal 17, 177
translocation 46, 54
transplantation surgery 63
triploidy (3N) 44, 46, 48, 54
trisomy 44, 49, 54
Trojan horse mechanism 148
tubulonephropathy 11
tumor 179
  malignant 63
tumor necrosis factor (TNF) 149

turkey 50

V

vasopressin (VP)
  ACTH release 61
  deamino-D-arginine vasopressin (DDAVP) 11, 158, 162
  hypophysiotropic effect 62
  precursor 10
  second messenger 12
vector
  retroviral 177
ventral tegmental area 99
virus
  caprine arthritis encephalitis (CAE) 147
  Maedi Visna (lentivirus, sheep) 147, 149
visceromotor reflex 60
vitamin D
  bone metabolism 127
  metabolism 126
  osteomalacic disorder 112
  rickets 112
VLDL
  see lipoprotein, very low density
VP
  see vasopressin

W

Western blotting 149
wild animals 66, 167

X

X-linked
  hemophilia 158
  hypophosphatemic rickets 112, 116
Xenopus 81
XLH
  see rickets, X-linked hypophosphatemic

Z

zoological garden 66
zygote 43, 44

Roy Mack

Dictionary for Veterinary Science and Biosciences
Wörterbuch für Veterinärmedizin und Biowissenschaften

German-English/English-German/With trilingual appendix: Latin terms
Deutsch-Englisch/Englisch-Deutsch/Mit einem dreisprachigen Anhang: Lateinische Begriffe
By R. Mack, Weybridge/England. 1987. 324 pages. Soft cover DM 49,80

This Dictionary has been compiled as a result of many years experience of translating texts in the biological sciences, particularly veterinary medicine. The aim is to supplement the general dictionaries with technical terms to in the fields of anatomy, microbiology, physiology, parasitology, pathology, pharmacology, toxicology and zootechnics, with special reference to domestic animals and their diseases. The commoner wild animals and those present in zoos are also included. As a specialized vocabulary the work is addressed to students, practicians and scientists in veterinary medicine and in the biological sciences; it is indispensible to professional translators, libraries and authorities.

Dieses Wörterbuch entstand aus der langjährigen Erfahrung des Autors bei der Übersetzung biologischer und insbesondere veterinärmedizinischer Texte, um die allgemeinen Wörterbücher mit dem speziellem Wortschatz aus den Gebieten Anatomie, Mikrobiologie, Physiologie, Parasitologie, Pathologie, Pharmakologie, Toxikologie und Zuchthygiene wirksam zu ergänzen. Auf Haustiere und ihre Erkrankungen wird dabei besonderer Wert gelegt. Darüber hinaus werden auch in die Europa häufiger vorkommenden Wildtiere und Zootiere berücksichtigt. Als Spezialwörterbuch wendet sich das Werk an die Studierenden, Praktiker und Wissenschaftler der biologischen Wissenschaften und der Veterinärmedizin; es ist unentbehrlich für Berufsübersetzer, Bibliotheken und Behörden.

Price: Status October 1987

Berlin and Hamburg